Our
Fated Century

Our *Fated* Century

Grant Rodkey

To order additional copies of this book, contact:
Xlibris
1-888-795-4274
www.Xlibris.com
Orders@Xlibris.com
751548

CONTENTS

Rodkey's Aphorisms

You can't learn nobody nuthin-the process requires some steam on the part of the recipient.

Do not look at the surface of the wound - look into the tissues and see important structures before you cut them.

The Law of the Blood Vessel: If you don't cut it, it won't bleed.

The eye is instinctively drawn to the jet of a bleeding vessel; train your eye to look behind the jet.

The term "First Assistant" is a misnomer-it should be "Manager of the Operation". If the Surgeon has to pause and think, the First Assistant has missed a cue.

You never saw a surgeon who was not gentle-he would have changed long before you met him; what you do see is an astonishing difference in the perception of gentleness among surgeons.

All operations should have a common objective: "Keep the strain off the macrophages".

All psychomotor learning follows the same pattern: careful analysis; slow, attentive , halting performance; and progressive accuracy, fluidity and speed as control moves to subcortical centers. This leads to the "Mary Lou Retton Phenomena"-a pretty, naive high-schooler picked up by a Hungarian gymnast –coach who knew exactly what she had to do: drilled her, drilled her and drilled her in correct performance until she could do it no other way.

Van Cliburn, the first American to win an International Piano Competition in Moscow, said, "I was fortunate to have had parents who were both professional pianists-I never practiced without expert supervision". It is much better to learn correctly and build on this than to learn, de-learn and re-learn. Corollary: By the end of the second year of surgical training a Resident becomes essentially un-teachable in terms of surgical technique. Be careful what you learn and when you learn it!

The body loves to be cut and to heal crossways-respect Langer's lines!

The anal sphincter is the prime muscle of human civilization-cherish and preserve it.

Everybody has to die of old age sometime; but does it have to be us or our patient during this operation?

Any structure which has survived 50 million years of genetic fine-tuning has a function; whether or not you know what it is.

A hog that is fattening ain't in it for luck.

The tears of gratitude dry quickly.

There is no person whom you will ever meet from whom you can not-and should not-learn something.

Boston VA HCS   August 10, 2003

Our Fated Century

This is a story of unique adventure—the transition (within one century) from postmedieval society (early or late, depending upon where on the globe you were born) to the space age and beyond. "Beyond," in this case, refers to the halting beginnings of our understanding of the unimaginable complexity of creation's most intricate achievement—*Homo sapiens.*

This is not my autobiography, although in the process of reviewing the cataclysmic changes in the world during my gift of the blessing of life, the reader will come to know many details of my own life, as well as the lives of my family and many friends who have guided and sustained me. Rather, it is an accounting of the remarkable events during our tenure, far more profound than man-made global warming, which have led me to the view that we are, indeed, in a fateful period in time.

We live now in an age which sees all people of the earth in a turmoil of uncertainty, anxiety, want, peril, pain, or misery. Even the Earth itself is convulsed by unwonted extremes of climate, storms, earthquakes, fires, and floods. Surely, this drama is not unfolding at the direction of man. In our search for answers, we must review our slide into this accelerating vortex. This was our fated century.

Chapter 1

In the Beginning

The beginning of my own life was long before my birthdate on November 17, 1917. The genes, attitudes, habits of thought, behavior, and moral values that formed me were hammered out through countless generations of struggle—of which I know only a few. On my father's side, our ancestors immigrated to Lancaster, Pennsylvania, from Southern Germany in 1795. Successive generations migrated westerly so that my grandfather (Grant Colfax Rodkey) was born in Indiana (son of Joseph Christian and Esther Dohner Rodkey) and moved with his family to Blue Rapids, Kansas. He married Kate Tyler, daughter of an itinerant preacher. My father, Joseph Verne Rodkey, was born in Blue Rapids, Kansas, in 1895, graduated from the Irving, Kansas High School in 1912, and attended one year at Kansas State Agricultural College in Manhattan. In 1914, the family moved to the high plains of Eastern Colorado to take ownership of a large cattle ranch twenty miles north of Limon.

My mother's ancestry included Mary Chilton, an adolescent (age thirteen) aboard the *Mayflower* in 1620, who married John Winslow (? in 1624) and is buried in the Winslow tomb in the graveyard of King's Chapel, Boston. Successions of pilgrims who migrated west included Wealtha Allen and Yardley Hough, who traveled by covered wagon from home in Heath, Massachusetts, to pioneer in Elkhart, Indiana. Their daughters—my grandmother Helen Elizabeth Hough and her sister Mary—were members of the first graduating class of Park College (Missouri). She became a schoolteacher where she met my English immigrant grandfather, Alford John Piper, who was a stone dresser for the water-powered flour mill, part-time postal transport driver, beekeeper, and horse stable (livery stable) owner in Irving, Kansas. My mother, one of six siblings, and her younger sister Caroline, walking, attended twelve years of school in Irving without a single day tardy or absent—surely a sign of the discipline and morality of the times!

Mother attended two years at the normal school in Emporia, Kansas. In 1916, she married my father and took up the life of ranching. In those days, this included laundry with a scrub board, cooking (including baking) on a kitchen range fueled by dried buffalo or cow chips, gardening without irrigation, canning in mason jars for the winter's supply, baking bread, shopping in Limon two times a month (on rare occasions traveling by team and wagon—a sixteen-hour expedition), and cooking for the family and the hired hands (cowboys). Mother was also an expert pianist, a trained vocalist, and a certified elementary school and music teacher. She had an Emerson upright piano on which she practiced faithfully every evening after bedding the children—an exercise that would rock the whole house and which I later described as "rocking around the house with Rachmaninoff!" She was particularly fond of Rachmaninoff, who was only twenty-two years her senior.

There were seven children born into the family. Following me in succession were Lee, 1919; John, 1921; George, 1922; Mary Jane (who died in infancy of congenital heart problems), 1928; Elizabeth, 1930; and Kathryn who was born in 1932. More of their lives will be brought into the story as we progress.

Chapter 2

Weaving the Tale

On the occasion of my ninety-fifth birthday, I resolved to record for my children (and those of others who may be interested) some items that I have learned and that may be interesting to them. Not that I expect them to learn from my advice—that would be contrary to my theory of education, which goes approximately as follows: "You can't learn nobody nothin'!" Without active input from the learner, nothing sticks.

Experience and observation have convinced me that there is a transparent, semipermeable membrane that separates generations. We can see through it and hear what is said. We can touch and feel and hug each other, but the membrane as barrier says, "No wisdom can come through here!" Each individual seems required to learn many lessons by his own (sometimes) painful experience before he can understand the meaning of advice given by someone who has walked the same path before him.

But I have lived a long, active, and observant life during a century that has changed the thinking and behavior of the world. Perhaps a nugget of description or insight here or there will strike a spark in the minds or hearts of my younger brothers and sisters to help them understand themselves or the world in a different light through the prism of my experience.

At the outset, I shall describe my general outlook and beliefs. As we proceed, the reader may understand how I came to hold these convictions and their relationships to my life.

First, I have absolute faith and belief in the power of the living God who created, and is still creating, the ever-changing universe and all its creatures, of which Man is the most highly endowed. Sir Isaac Newton, in his *Philosophæ naturalis principia mathematica* (1687), expressed his shocking conclusion that "space is the consciousness of God." This, I believe and all that is within that sphere is his creation—a complexity unthinkable to man.

But God also is a present power and interested party within each one of us. From the moment of our conception when that spark of life is transmitted to us, the unbroken chain from His creation—the living, pulsing power of God through the DNA in each cell of our structure—has directed our form, development, function, and thought. Regardless of friends, family, beloved companions, crowds, or loneliness, each of us walks the pilgrimage of life alone in his deepest recess. There is an inescapable emptiness—my hollow spot—deep within each of us that yearns for the companionship of God. It has been said that "we come into the world with nothing, and we leave it with nothing." How wrong! We come into the world with the fire of the living God surging in our veins, and we leave in the arms of his mercy and love!

The psalmist has said it best:

> O Lord, thou hast searched me and known me. Thou knowest my downsitting and mine uprising, thou understandest my thought afar off. Whither shall I go from thy spirit? Or whither shall I flee from thy presence? If I ascend up into heaven, thou art there: if I make my bed in hell, behold, thou art there. If I take the wings of the morning and dwell in the uttermost parts of the sea; Even there shall thy hand lead me, and thy right hand shall hold me. If I say, Surely the darkness shall cover me; even the night shall be light about me. Yea, the darkness hideth not from thee; but the night shineth as the day; the darkness and the light are both alike to thee. I will praise thee; for I am fearfully and wonderfully made: marvelous are thy works; and that my soul knoweth right well. Search me, O God, and know my heart: try me, and know my thoughts: And see if there be any wicked way in me, and lead me in the way everlasting. (Psalm 139)

Chapter 3

WHO AM I?

Each of us needs to find the answer to that question. We need to know where we stand in the family, in the school, church, and the community at large. Sometimes, we may even speculate whether we really are whom we seem to be!

Early Years

In my own case, I was born at seven o'clock on a Saturday morning on November 17, 1917, in my grandmother Kate Rodkey's house in the small prairie town of Limon, Colorado. Both physicians in Limon (Drs. Kessinger and Kennedy) were in attendance, and the delivery was difficult, aided by forceps that dented my head—no permanent damage that I know! My parents, Helen Elizabeth Piper and Joseph Verne Rodkey were both age twenty-two and had been married for a scant fifteen months.

As the firstborn, I received great doting from my parents, grandparents, and three teenage aunts—Gladys, Margaret, and Anna May. I was named Grant V. Rodkey after my grandfather, Grant Colfax Rodkey (so-named because he was born in 1868 when Ulysses S. Grant and Schuyler Colfax were elected as president and vice president). The *V* was a token to my father who was called Verne—as neither he nor Mother liked "junior." That *V* has caused me more than a little trouble—first, because people always want to know what horrible middle name I really have that I won't disclose it; second, because important documents want to have the entire name spelled out—or a plausible explanation (initial only).

Finally, when I got to Harvard in 1939, they asked what the *V.* stood for. When they heard that it was an initial only, they said, "Well, then you can't have a period."

"But I have always had a period!"

"No matter, you can't have it here."

And on my graduation diploma in 1943, surely enough, there was engraved the *V* with no period. However, over the course of seventy years, I have beaten Harvard down. I am now usually addressed with my period!

Although I was born in Limon, our home was on a ranch twenty miles north of the town and five miles from the nearest neighbor. At that time, the population in Colorado was four per square mile—and most people lived in Denver! We had few amenities: an outhouse at a suitable distance from the dwelling, a kitchen range and a living room stove which burned (mainly) cow chips because wood and coal were not available, and a crank telephone—sometimes functional—with a party line shared by twelve other families. There was no electricity, and night illumination was by kerosene lamp or lantern. By virtue of a windmill and an elevated tank, we had running water to one spigot in the house and to two stock tanks for the cattle and horses to drink. We sometimes had a Model T Ford car, but transport was chiefly by horseback or by team and wagon. Food was mainly what we raised on the farm and preserved by mason jar canning or in a root cellar. Meat, milk, and eggs were homegrown; butter was churned from cream in a crock jar with a wooden paddle. And pickles were cucumbers preserved in brine. Sugar, flour, salt, and other staples were hauled by team and wagon from Limon—a trip that began at 3:00 AM and took until 10:00 or 11:00 PM that night. Laundry was tub-and-scrub-board technique, and there was no Dy-Dee diaper service! Radio had not been invented.

But we had a piano! It was a wonderful piano—an Emerson upright, baby sister of the Steinway—with excellent tone and outstanding keyboard action. Both my parents were musicians—Dad with his singing and Mother both vocal and instrumental with her piano. She was, in fact, at the stature of a concert pianist, as well as a trained elementary school teacher. So our home was always ringing with music—classical as well as popular. I remember vividly being carried upstairs to my crib which was directly above the piano; after which, Mother would sit down for two hours of practice—scales, arpeggios, Chopin, Mozart, Beethoven, and Rachmaninoff. She was particularly fond of Rachmaninoff (who was only twenty-two years older than she) and played his *Prelude in C-Sharp Minor* with great gusto, shaking the piano and the whole house! In retrospect, I have called that "rocking around the house with Rachmaninoff!"

Aunt Margaret was also a very gifted piano player, and she had a talent that I have never seen in another person. She could listen once to any piece, popular or classical, then sit down and play it in its entirety—melody, chords, and all! And remember it thereafter! Unfortunately, it was too easy for her, and she was never willing to concentrate and practice as Mother did; therefore, she made less of her talent.

Fig. 1: (1918) Rodkey Ranch building complex-stockyards and barns against the horizon. Grant V. in the stroller w/Aunts Margaret and Anna May.

Thus, our home was constantly alive with music: whistling, humming, singing at work, singing in groups. I could barely talk when I began to sing, and I had my stage debut in the Limon movie theater at age four. I remember the evening as if it were yesterday. There were electric lights, a stage, a piano below the stage where a pianist inserted *ad lib* sound effects for the silent (black-and-white) movies, and a crowd—the largest group I had ever seen—perhaps fifty to seventy-five people! I was terrified, and so I asked Mother if I had to sing.

"Yes, son. You have to sing."

Boy! Ms. Margaret Thatcher has been dubbed the Iron Maiden, but my mother had a lock on that appellation long before Ms. Thatcher hove onto the scene! So I sang a little song about "sweet Peggy O'Neal" without forgetting the words—a marvel! And then in response to rave applause, this one-line encore: "Robin-a-Bobbin bent his bow; shot at a pigeon and killed a crow!"

Of course, by this time, I could reach the piano keyboard, so my fate was sealed! Parenthetically, I forgot to mention that while I was crawling around on the floor in my infancy, the hems of women's skirts (always black!) struck me just at eye level. Very intriguing for a kid of my stature! This was before the women's suffrage amendment to the Constitution and, certainly, before the age of liberated women! In fact, in those days, they were all corseted!

However, my life was surely not confined to indoors. When I was three, my father gave me a saddle horse who also was three; and for many years, Valentine had the more sense of the two of us! In fact, looking back, I realize that Valentine took very thoughtful care of me through many unreasonable experiences of my own initiative. The only way I could clamber aboard was to lead him to a fence post, then climb up the fence wires, and jump across onto his back. Never once did he ever shy away and make me miss! We traveled over the countryside together with great happiness. Very quickly, I took up the job of bringing in the cows for milking in the evening. Valentine was expert at that too. He knew just which of the

7

cows were to be brought in and which were to stay in the pasture. And he was a whiz at cutting off some animal who might wish to bolt and escape! In the summer on the treeless plains, the sun bore down on us fiercely. However, occasionally, a cloud would float by overhead, providing heavenly shade. Valentine and I would often head for the shade and ride along with the cloud for a time to cool off. Once, my father called me in to say that the neighbors complained that I was speeding on horseback!

On one memorable occasion, my brother Lee had ridden off on Valentine midmorning but failed to return as expected. About 2:00 PM, Valentine appeared at the back door of the house, neighing, stamping his feet, and shaking his bridle. We said, "Where is Lee?" Whereupon, Valentine wheeled around, led us out into the prairie for three miles where we found Lee walking around, lost. Lee had no memory of events, so the horse must have fallen, thrown him so that he had a concussion then come back to the house to summon help!

Riding across the plains, we saw many jackrabbits, prairie dogs, chipmunks, rattlesnakes, badgers, coyotes, skunks, antelopes, and occasional whitening skulls of buffalo and many flint Indian arrow heads! Meadowlarks, bobolinks, prairie owls, hawks, bald eagles and night hawks, killdeer, and occasional seagulls (probably from the Great Salt Lake in Utah) were our companions. The green buffalo grass carpeted the prairie and its low, rolling hills, turning to brown in the fall. The atmosphere was clear—so clear that we could see Pikes Peak one hundred miles away and identify timber line on its eastern slope. There was always plenty of fresh air and, always, sunshine! In those days, hotels in Colorado gave free lodging to their guests on any day that the sun did not appear for some part of the day!

When I was seven, we had a great three-day blizzard with heavy snow and furious, bitter winds. On the morning after the storm passed, Valentine and I set out to find and round up the cattle that had drifted in the storm. There were few fences at that time. As we rode along, the sun shone brilliantly on the snow on the plains—so much so that I got a bit of "snow burn." But we found many small animals and birds sitting in their nests or lairs—all frozen to death! Some of the cattle had drifted eight miles, but we found and rounded them all up. About noontime, we passed the Middlemists' ranch, and they insisted that I come in to have some lunch. However, I was too shy to get off my horse, so they brought me a sandwich! That family had twin sons, Robert and Louis, about fourteen years old. Louis had diabetes and was quite ill. Just that year (1924), Banting and Best in Toronto discovered insulin, so Louis was spared to live a long life in Denver.

I was the first but not the only child in the family. Lee came along in 1919, John in 1921, and George in 1922. We were a purely masculine outfit until Mary Jane was born in 1928. However, she was a cyanotic baby (? congenital heart defect) and lived only ten days. Dad was magnificent in helping the family through this crisis and in attempting to teach us something of the meaning of death. Then Elizabeth came along in 1930 and Kathryn in 1932. Each child in turn took up singing, the piano, chores, hoeing in the garden, and helping with the animals (horses, cows, pigs, sheep, goats, chickens, and turkeys). At age eight, I was driving a four-horse team, harrowing or disking in the fields. And the team included old Charlie, the loco horse whom you could never trust! At age ten, I was breaking sod with a Fordson tractor and two-bottom plow. Each member of the family contributed as much as possible within the limits of age and ability. Granddaddy Rodkey also worked with us in the fields. These were full and busy days, and happy days.

In particular, I recall that I (age six and a half, barefoot), on a hot, sunny July afternoon, carried a one-gallon covered bucket of water to Dad, Granddaddy, and Uncle Ira Tyler (Grandmother Kate's youngest brother) who were harvesting wheat with a header in a field 1 x ½ mile in dimension. They were working in the middle of this tract. (A header is a machine powered by a four-horse team, having a platform with a cutting cycle at its leading edge. It had a revolving flat canvas platform which transported the grain and stems to the left end of the platform where it was picked up by an elevator canvas and pitched into an accompanying hay wagon.)

The men stopped the rigs, wiped their brows, and gratefully drank the entire gallon of water! But as they were paused, we heard a strange buzzing sound. Looking up, I saw my very first airplane! It looked to me like two ladders crossed at right angles! I got no insight into the third dimension! In those early days, planes were made with wooden frames, and what appeared like rungs of a ladder were braces supporting the frames of body and wings!

The conversation was about the coming election: Calvin Coolidge as presidential candidate and Charles Curtis as vice presidential candidate. Not much was known about Coolidge, but Curtis was a Kansas man and certainly okay! This was the summer of 1924.

When I was six, I was unable to attend school—there was none! The state law in Colorado required that there must be at least three children of school age within a school district for the district to hire a teacher—a district being about ten miles square. That year, there were only two—Allen Houston and me. So we had to wait until the next year (when Lee became six) to have a school. Meantime, Mother taught me to read, and I took to it readily. In the spring of 1924, Lee was six years old, and we reached the magic number! School opened, although it was about two-and-a-half miles away, and we had to go on horseback. It was my job to tend to the horses with feed and water at school. We stored hay for the horses and coal for the school stove in a small shed placed between the boys' and girls' outhouses.

And so we began—with Allen Houston in the sixth grade, Lee in the first, and me in the third—although in midyear, I advanced to the fourth grade and completed that as well. The school was one large room with a small anteroom for coats and boots. The stove stood in the middle of the aisle two-thirds down toward the front. There was a teacher's desk in the front, a blackboard behind her, a dusty pump organ at the side, and a row of students' desks down each side of the central aisle. Recitations and assignments of all students were, of course, heard by all the others—a significant learning process in itself! Of course, we all carried lunches. Recesses were observed out-of-doors with a combination of boys exploring nature and playing marbles, duck (rock)-on-a-can, footracing, or abbreviated baseball.

We played baseball with special rules: we had a pitcher, a batter, and a catcher. If the hitter didn't make a home run, he was automatically out! Hide-and-seek was not much of an option. There was no place to hide on the open prairie!

Our teacher was Evelyn Holsclaw, a young woman from Otis, Colorado, who had one year of experience teaching at another school in northern Colorado. The only feasible place for her to live was in our home, and of course, she also rode horseback to and from school. It was her job to build a fire and keep the school warm—sometimes difficult—as well as to teach and monitor her charges. It was a particularly fortunate opportunity for us since she was a gifted teacher and a lovely person. It was a year of great happiness, and in its course, I fell deeply in love with Evelyn to the point of being very jealous of John Middlemist, who could squire her around in his Willett sedan, and her fiancé, Harold Janke, whom I rarely saw but who won out in the end. When she was married, I chalked that down as a black day, which I remember each year to the present time! I kept in touch with her until her death in 2011 at age 106—a lifelong love affair!

But civilization was coming! Calvin Coolidge had been elected vice president in 1920 and succeeded to the presidency when Warren Harding died in 1923. He was elected president in 1924. In those days, the new president took office on March 4 of the year following the election. On March 4, 1925, President Coolidge was sworn in and gave an inaugural speech over the radio! The Middlemist family had bought a radio—the first in the whole north-of-Limon territory. On that March 4, the neighbors from miles around came by team and wagon to spend the day at Middlemists', listen to the radio and President Coolidge, and visit. This was my introduction to the communications media!

As was the case then in all community gatherings, talk soon turned to the events and the significance of the Civil War. At that time, there was no family who had not been affected by this cataclysmic event—wounding, suffering, death, family separation, or alienation. There was, as yet, no Memorial Day. It was Decoration Day to decorate the graves of the fallen soldiers from those sad years. (It is a significant fact that in December 2014, the United States government was still paying a monthly pension of $73.13 to Ms. Irene Triplett, daughter of Mose

Triplett of Wilkes County, North Carolina, who was born February 4, 1846, fought in and survived the Civil War, and died July 18, 1938. The scars of war heal slowly!)

But we did have Arbor Day when each child was taught to plant a tree—although there was pretty scant chance of it ever surviving on the arid prairie!

During the summer of 1927 came the first of several critical events that were to direct my life. On a sunny June day, I was doing patch repair work on the water storage tank that was mentioned earlier. The tank sat on an eight-foot high base with a narrow catwalk around its edge. One of the boards of the catwalk had become rotten and gave way, pitching me eight feet to the ground into a pile of rubble. When I got up, my right arm was useless, with the lower end of the humerus sticking out through the skin of my elbow and covered with dirt, and my wrist was badly deformed. In addition, with each heartbeat, a jet of blood shot out eighteen inches from my elbow. I began screaming and ran toward the back porch of the house, which was about fifty feet away.

Dad was in from the field for lunch—an unusual event since he always carried lunch with him. He heard my screams and came running to meet me on the back porch. He immediately put a tourniquet around my arm, stopped the bleeding, wrapped my arm in a towel, and laid me on the living room floor. Fortunately, Dr. Kessenger was in his office in Limon, and he came at once. I sang the words to every song I knew to relieve the severe ischemic pain in my arm during the forty-five minutes it took him to arrive!

Dr. Kessenger put me to sleep with drop ether on the floor where I lay, then he gave the can to Dad to continue the drip while he (Dr. Kessenger) worked with the arm. He confirmed a compound dislocation of the elbow with severe contamination and compound fracture of both bones of the forearm about two inches above the wrist. The brachial artery had been transected at the elbow, but the proximal end had retracted and thrombosed so that there was no bleeding when the tourniquet was released. When the wounds had been cleaned, the dislocation reduced, and the forearm fractures set and splinted with plaster of Paris. I was allowed to awaken. I remember vividly my first reaction: "My, what a restful sleep!" And the pain was remarkably lessened!

We then got into Dr. Kessenger's car and drove to the nearest x-ray—thirty-five miles away in Hugo. The first films showed perfect alignment, but the physician who took the films was rough in handling me. While we waited for film development and interpretation, my arm began hurting badly again. Dr. Kessenger was pacing the floor, and when the x-rays were brought in showing the spectacularly good result, he demanded that new x-rays be taken. Surely enough, the forearm fractures had been displaced by the rough handling! In great disgust and anger, Dr. Kessenger put me back into his car, drove the fifteen miles to his office where he reset the fractures, and applied a cast from shoulder to wrist and waist without benefit of anesthesia or x-ray! He left a cut-out window over the elbow wound for dressings and sent me home with Dad.

Each day for the next six weeks, my parents drove me to Limon for Dr. Kessenger to change the dressings on my wounds. This was before any of the antibiotics had been developed, but Dakin's solution, a 0.5% solution of sodium hypochlorite in water, had been proposed and used as an antiseptic during World War I by Dr. Henry Dakin and Dr. Alexis Carrel. Its antiseptic action is related to release of nascent chlorine. Daily dressings with Dakin's solution allowed my wounds to heal without infection—even in the face of gross contamination at the time of injury. Of course, this was without antibiotics—none of these had had been discovered then!

My arm remained in a cast for three months, and it took two years to get most of my motion back (the ability to get a fork to my mouth with the right hand!). But eventually, I developed a normal range of all motions with no deformity or weakness, and I even regained a radial pulse! This is a result that would be difficult to attain today with all modern technology and drugs. My life was forever changed by the skill and devotion of J. D. Kessenger, MD, a railroad surgeon at Limon, Colorado.

For the first two or three days after my fall, I was quite groggy, but then I realized what a blessing in disguise had befallen me! Mother's piano students were all preparing for a recital; and each was required to learn, memorize, and play a musical selection appropriate to his/her progress in learning. As one of those pupils, I realized that there was no possible way for me to do that with a full-length arm cast. Hot diggity dog! I was off the hook! But I made the mistake of pointing this out to Mother. "Oh no, you don't get out of it! I have a left-handed arrangement of *Believe Me if All Those Endearing Young Charms*. You will memorize and play that!" So not only did I have to play, but also I had to memorize an entirely new composition—one I remember to this day, the *Iron Maiden*!

Of course, there are many other things that a right-handed person does as a matter of course. Eating, dressing, bodily toilet, writing, and countless other usual activities had to be transferred to my left hand for nearly two years until I could get a fork to my mouth using my right hand. As a result, I became essentially ambidextrous. This experience has been reflected subsequently in my training of left-handed surgical residents whom I always encourage to become ambidextrous. I advised them to switch deliberately to their right hands in eating and dressing and other frequently performed activities to convert what is an inconvenient, moderate handicap to an advantage.

Before my injury, I was showing some promise as a baseball pitcher. Dad, who was a semipro player, knew a remarkable amount about the tricks of pitchers—the same ones that they use today—and he was well along in coaching me. However, afterward, I was never able to throw a ball hard without developing pain in my elbow—a very small price for a miracle!

My remaining grade school years were completed in a one-room school but with more students and more grades represented. During those years, I observed that younger students learn a great deal from older students, and the latter benefit from being assigned to help the children in earlier grades. The arrangement was certainly no hardship for us. During those years, I read voraciously, including such works as Shakespeare's plays, *Ivanhoe* by Sir Walter Scott, and *Les Miserables* by Victor Hugo. We also had a set of *Encyclopedia Britannica* that was my constant reference. Most of the reading of literature and reading for pleasure in my whole life was during this period, since during later times, I was always working to support myself and to defray my educational costs—as well as studying.

Chapter 4

HIGH SCHOOL DUST AND DEPRESSION

In the fall of my twelfth year, I was ready for high school. This meant "Limon"—too far to drive and too expensive to pay for board and room. So my parents arranged for me to have a job for board and room on the Morrissey farm just three miles north of town and on the school bus route. I began a new pattern of life: up at five in the morning to milk nine or ten cows, separate the milk into cream and skim milk, feed the skim milk to the calves and pigs, and be ready for the school bus at 8:00 AM. Because of peculiar routing to pick up other students, we went nine miles before we reached the school! After classes and reversing the bus route, I would ride out to the pasture on horseback, bring in the cows, and repeat the routine. Study later. Saturdays: work in the fields, picking corn or binding or shocking grain. Milking and other chores were a seven-day/week routine!

The Morrisseys had a daughter in high school, but I saw very little of her—aside from the bus rides! There was one memorable bus ride when, as we were headed home, a very sudden and fierce snowstorm set in. Within five minutes, we could see *nothing*! Fortunately, we were within a half mile of a house, so we all got out of the bus, holding hands in a chain. We found our way to the roadside fence (barbed wire!) and held onto each other and to the fence until we reached the house, where we were marooned for three days. Luckily, the family had some stocks of food and a telephone so that our parents knew we were safe!

Oddly, I remember very little of my classes in high school—geometry, algebra, Latin, history, English, and band (trumpet). Mainly, I remember that Doris Nims always seemed smarter than me, and she ended as valedictorian, and that Floreine Anderson was the prettiest girl in the school! Somehow I wound up in track and groveled around in wrestling! Of my teachers, my bandmaster, Mr. Raymond Hunt, and our superintendent, Mr. Smith—a burly graduate of the University of Missouri in Columbia—are the only ones who come clearly to mind

14

after eighty years. I do remember playing my trumpet in a Saturday night dance band on some occasions! My favorite tune: "Stormy Weather"!

Most of the important events of those years happened outside the school. Lee and the younger children were approaching time for high schooling, so Dad and Mother decided to move nearer to Limon. We rented the Maynard farm three miles southeast of Limon—alongside the Union Pacific Railway, adjacent to Lake Station (a watering station for the trains' steam engines).

The migration of family, machines, and animals from the old ranch was a difficult emotional and a tough physical event, but was accomplished during the summer of 1931. This put us children on a school bus line, and we all entered Limon schools, high school, and grade. Coming out of the isolation of country life, we were all exposed to the various infectious diseases of the urban population which we brought home successively, and they then ran through each of the family members successively—mumps, measles, and scarlet fever. In those days, quarantine was imposed for those diseases. That is, you were required to stay home so as to avoid exposing other people to your diseases. Our family was quarantined the whole spring of 1932, and we were at times very ill. It is hard to realize, in retrospect, that this was before any antibiotics were known. It certainly is a marvel that we all came through.

However, not all bad news was within our family. What has been named in retrospect the Great Depression was ushered in by the Wall Street stock market crash of 1929. Many—if not most—banks failed, people lost their savings, businesses closed, and people were thrown out of work by hundreds of thousands. Bills that people had accumulated had to be paid regardless. And if you didn't have the money to pay off, the creditors seized your property and auctioned it off. That happened to us, and they auctioned off all our farm implements, our animals, and anything else of value in one terrible day. But there was a redeeming event. Very early on the morning of the auction day, our best team—Buster and Belle—were harnessed and hitched to a wagon, and Valentine was saddled up. Dad sent Lee to take the brigade up over a big hill to the far side of the pasture to tie them to a fence until all the vultures had left. And we saved the piano. Thus, we saved a remnant!

Without animals, we could no longer farm, so we moved to a house with a small barn on the south side of Limon—from which we could walk to school. For a time, Dad had a job with the Works Progress Administration (WPA) as part of President Roosevelt's attempts to infuse some money into the economy. But taking care of basic necessities of life for the family was a tough job for our parents. At intervals, Dad would drive a wagon along the railway right-of-way, and we boys would scour the strip adjacent to the tracks on either side to scavenge coal which might have fallen out as the fireman scooped fuel into the steam engine boiler. It goes without saying that no opportunity to work and make a buck was ever overlooked. Among other efforts, Mother continued teaching piano students, and that activity brought in a little money or bartered food.

And then there were the dust storms! Beginning in November 1932, intermittently, huge clouds of dust would roll in on high winds and smother the sun so that automobile drivers were forced to use headlights in daylight and creep along, peering to see the shoulder of the road over the right front fender of the car. Dust would blow right through the houses in a fine mist that made the electric light (in the middle of the room) look dim. And the dust piled up in drifts like snow on the leeward side of any obstruction like a piece of machinery or a fence post. When I visited that country in 1964, those dirt drifts were still clearly visible, and in 2011, they're still identifiable! This was called the Dust Bowl.

In order to understand the Dust Bowl, one needs to know a little history. Until 1917, that plain's country had all been grazing range huge ranches with herds of cattle. For instance, my dad and grandfather had a ranch which included sixteen sections (sixteen square miles or 10,240 acres) together with approximately four to five thousand heads of cattle-and the horses (and cowboys) to manage them. However, when the United States entered World War I in 1917, the government anticipated that we would have a great need for increased food production. The ranchers were encouraged to take out Federal Reserve loans, go to Texas and buy large numbers of young cattle, and bring them up north and fatten them on the range. Of course, everyone went for it—that was the patriotic thing to do, and you might make a buck too! However, the war ended in 1918, perhaps sooner than expected. And in 1919, in a single year, the government called in all those loans. Of course, the cattle were the only "liquid assets" that ranchers had, so they all had to sell off stock to pay up. Dumping all those cattle on the market at once depressed the price of beef so severely that everyone was bankrupt—everyone, that is, except my dad who would never take out bankruptcy papers and eventually paid off an agreed fraction on his debts to all his creditors.

But the only option left to the ranchers was to lease back their buildings and some surrounding land to do dry farming. Eastern Colorado is just one notch above a desert, which is defined as having less than ten inches of rainfall per year. We averaged fifteen inches per year. But from 1928 onward, it was so dry that we never had a crop. Although we tilled and seeded and cultivated, the land crops simply withered away. And the land became progressively dryer. Grass on the plains has very deep and complex roots which hold the soil in place; but plowed soil has no such protection. In due course, the high winds began to lift and blow away this topsoil in literal seas of dust.

In the midst of all this turmoil, I graduated from high school in late May of 1934—Doris Nims the valedictorian, I the salutatorian, and some twenty-eight other classmates following. One of the benefits of my achievement was a first-year scholarship to the University of Colorado in Boulder that I never cashed in!

Our family had recognized that we were coming to a dead end. For several years, our father had been turning the issue of resettling in his mind. Aunt Gladys; her husband Julian Howe; and their children—Doris, Herbert, and Orville—had been driven from Colorado by the fierce economic slump in 1932. They had moved to a small farm in Stephentown, New York. Dad had questioned closely those who traveled to the Pacific Northwest, and in 1933, our parents determined to migrate to that area. During that year, Granddaddy, Grandmother Kate, Aunt Margaret, and her husband Royal Westphal had driven to northern Idaho and found small irrigated farms near Post Falls. They bought two of these and leased the third one for us when we should come at the close of the school year. Thus, the stage was set for change!

Chapter 5

THE IDAHO MIGRATION

The first phase of the Idaho move—the 1933 beachhead—had been well-organized, and we had a place to which we would move in Post Falls. With school closing at the end of May 1934, we were ready to pull up stakes and leave, but what must we move in order to function in our new home? We needed farm equipment, horses, wagon, plow, tools, household furniture, and of course, the piano! So it was decided that we would hire an "immigrant car"—a railroad boxcar that we should fill with our belongings then hook into a freight train to be hauled to Idaho, although the rail destination was to be Spokane, Washington, some twenty-five miles west of Post Falls.

Uncle Royal worked for the Union Pacific, so he was very knowledgeable about such matters. And his wife, Aunt Margaret, was entitled to a free rail pass for passenger travel. I don't know what that boxcar rental fee was, but I know it took all the money the entire family could scrape together with barely enough left over to pay the cost of driving our family car (a 1928 Chevrolet six sedan) to Idaho with its precious cargo—Dad, Mother, Lee, John, George, and Kathryn, now two years old. Aunt Margaret was to go by passenger train on her rail pass and take Elizabeth, now three and a half.

That left Uncle Royal and me. We were to travel in the immigrant car. In fact, we were entitled to have one passenger in the car to water and feed the horses. But we had no other way to get me transported, so we packed the load, leaving space just large enough for me to lie down in the forward end of the car, which I had to reach by clambering over the piled furniture and equipment that reached from the floor to the ceiling of the car. In addition, we had Buster and Belle with their food for the trip, as well as the wagon, the plow, and Uncle Royal's car. So we were packed to the gunwales!

On Thursday evening following school closing, we had a wonderful farewell church service and party in the Limon Methodist Church. It was a deeply emotional experience, for we were leaving the only friends whom we had ever known and with whom we had shared the hard and bitter life of pioneers in a harsh land. And we were going into who knows what! When we sang the great hymn "God Be with You till We Meet Again" there wasn't a dry eye in the place. We all knew that the only place we were likely to meet again would be in heaven!

Early on Friday morning, I crawled up and over into my hideaway, and we loaded the horses. At the last minute, Uncle Royal took his place at the open boxcar left side door, and we felt and heard a great *BANG!* as the train backed hard into our car to engage the lock then began chugging off toward Idaho. Dad loaded the family with luggage and sandwiches into the car and pulled onto the highway with road maps at the ready. And Aunt Margaret and Elizabeth boarded a westbound passenger train for our common destination. We had left our home.

On the Road

The rail route took us to Denver then Cheyenne and Sheridan, Wyoming. In Montana, we hit Billings, Great Falls, Glacier Park, Kalispell, and Libby. Finally, we came down through Bonners Ferry and Sandpoint in Idaho and then to Spokane. By the time one straightens out all the curves in the track, it would run 1,500 to 2,000 miles, and it took us seven full days and nights on the way.

From Denver north, we followed the front range of dramatic high mountains rising out of the plains. Through Wyoming, more plains until we hit the Bighorn Mountains in northern Wyoming. From there, it was on through the high plains of eastern Montana until we crossed the Missouri River at Great Falls. From there, we climbed steadily into the Rocky Mountains, crossing the Continental Divide near Glacier National Park, then through various ranges of mountains until we came down into northern Idaho.

The scenery was spectacular and vast! Uncle Royal and I spent many hours watching prairie dogs, jackrabbits, coyotes, and antelopes out the open side door of the car as the miles clicked away under us. But the trip was also tiring. At night, we were invariably shunted off into some rail yard accompanied by bumping and crashing into our car as they made up the train for the next division run. We were somewhere in the middle of a very long freight train—over a mile in all. Of course, food was pretty monotonous—only dry and cold foods which would not spoil without refrigeration—with water from a jug that Uncle Royal was able to fill at stops along the way. As for me, I made myself scarce. If I were to have been discovered, we would have been required to pay a passenger fare from Limon to Spokane—an utter impossibility! But the constant interruptions of sleep by the banging around made us both very tired. Uncle Royal was becoming apprehensive. He sensed that the train crews had an idea that he might have a stowaway on board, so we became increasingly wary. The stage was being set for another miracle in my life.

As we came through the western Montana Rockies and were on the last day's run into Spokane, we debated the idea of having me get off the train at a time it was moving slowly, drop back and catch on again, climb onto the top of the train and ride there until we reached Spokane, then drop off and find Uncle Royal in the boxcar when the coast was clear. With the plan set, we waited for the opportunity when the train should slow.

We had this huge, long train (approximately one mile), which was being pulled by two mountain engines—big steam engines—each with sixteen driving wheels. So the train was taking the uphill grades pretty briskly. We studied the speeds, agreed that I should jump off facing forward, running as hard as I could to avoid being thrown down. At a certain point, with the train going uphill and on a curve to the right, it seemed as if I might be able to pull it off by jumping out our door, which opened on the left side of the car. So I jumped! Facing forward and hitting the ground running at top speed, I was thrown hard to the ground and over the shoulder of the rail bed and into the ditch. We had badly misjudged the speed of the train, which must have been going between forty and forty-five miles an hour. Fortunately, I was not hurt, and I picked myself up quickly just in time to see Uncle Royal disappear around the bend in the track!

I knew that I had to get back onto that train, no matter what. So I climbed back up onto the rail-bed, ran as hard as I could alongside the train and grabbed the ladder at the back end of one of the boxcars. It felt as if my shoulder would be pulled out of its socket, and my body was jerked right out straight in line with my arm and with my feet swinging up hard against the car behind me. But I held on! Gradually, I was able to pull my body forward and get a foot on the bottom rung of the ladder, then I climbed to the top of the car and lay down, holding onto the narrow catwalk running along the top of the train. This was no picnic. But we had not known about the *tunnels!* Of a sudden, the train headed into a tunnel through a mountain! I spread out flat, held on tightly, turned my head to the side, and went through several dark and smoke-filled tunnels, but none of the tunnels we passed through was low enough to scrape me off! Another miracle!

Finally, we emerged from the western edge of the high Rocky Mountain range, crossed the Montana-Idaho border, and began the slow descent along the bed of the beautiful mountain stream named the Kootenai River—a path trod earlier by Lewis and Clark. The scenery was beautiful, the vegetation lush, the scent of the pine trees exciting—a foretaste of heaven on earth! I had a ringside view of the whole panorama!

As evening approached, we rolled into the freight yards at Hillyard, on the north edge of Spokane—the end of the journey at last! As we had planned, I crawled down, walked along until I spotted Uncle Royal's car, then waited while it was shunted off onto a siding and the train crew had gone off. When I came up to the car door, Uncle Royal was the most relieved and happiest man on the continent! All he could think of was me in a cotton shirt and pants out alone in the Montana Rockies—grizzly bear country—miles from anywhere, perhaps injured or worse, and his responsibility! Certainly, another miracle!

We fed the horses, closed the doors and headed for Post Falls. All Uncle Royal knew was that there was a bus line to Post Falls on Trent Street. So we started walking to Trent Street, which turned out to be six or seven miles from Hillyard! But we did find a bus, went to the stone church two miles west of Post Falls, and walked the mile to Granddaddy's house, there to be reunited with all our family who had arrived a day or two earlier. What a happy time of blessing! The trip in the old Chevy had been uncomfortable, dangerous, and filled with anxiety, but all had made it safely.

Dad, Granddaddy, and Uncle Royal were very concerned that the immigrant car might exceed the prescribed weight limit. The car was due to be weighed the next morning. If it should weigh more than the agreed-upon limit, we would have to pay an additional amount of money before we would be allowed to unload our property—an impossibility since our money was in very short supply; that is, none. It was decided that we would go down and take as many items off the car that night as we could haul in Granddaddy's Model A Ford truck and a four-wheeled trailer, which Uncle Royal had bought a year earlier.

And so we set off—Dad, Granddaddy, and Uncle Royal in Dad's car with the trailer and Lee and I in the truck following—the trip being about an hour each way. We found the boxcar, unloaded all we could carry, and headed home. Lee was sleeping in the seat beside me, and I was fighting madly to stay awake.

Just after we crossed over the Washington State line into Idaho, the road took a curve to the left and then a long curve to the right. Somewhere on that long curve, I dozed off to be awakened as my left front wheel drifted over the shoulder of the road. Immediately ahead of me was an eight-foot boulder! I spun the wheel to the right with all my might, and the truck crashed sideways into the rock and then ran back up onto the road. Everything still seemed to work, and I got out to inspect. I found no damage except that the left running board had been dented upward! We were able to repair this dent with a sledgehammer the next morning!

Sitting directly behind the cab—with Lee and me immediately in front of it—was Mother's old gasoline-powered Maytag washing machine. It stood on legs about three feet tall and had a heavy aluminum tub just behind our heads. If we had hit head-on into the boulder, that machine would have come crashing through onto us, and we almost certainly would have been killed. Surely, another miracle!

Chapter 6

A New Life

We were completely unprepared for our new environment—green grass, trees, lakes, rivers, forests, and mountains! The small, irrigated farm which had been rented for us fronted on the bank of the Spokane River, and along the bank ran a large irrigation ditch—about ten feet wide and four feet deep—carrying water from the Post Falls Dam to Green Acres Farms in Eastern Washington. This was perfect for swimming, so all of us children learned to swim, which we could not do in eastern Colorado because there were no lakes or rivers. This was long before the days of backyard swimming pools!

On the first Sunday, we went to church at the Community Presbyterian Church in Post Falls to give thanks for our safe deliverance and to begin new friendships. These friendships were to outlast a lifetime, passing on to successive generations. The Stewarts, the Lindbergs, the Websters, the Moshers, the Eisenhauers, the Kildows, the Downings, the McQuowns, the Chapins, the Seyforths, the Wetherells, the McFarlands, the Bruggers, the Johnsons, and many other families took us into their hearts and lives with generosity and enthusiasm. We were finding our new home!

Transferring farming skills from over four hundred acres of dry farm to a small, intensively cultivated, irrigated farm was a real switch! That became my job, as Dad, fortunately, got a job as a section hand on the railroad to bring in a little money. I quickly adapted and taught the new ways to my younger brothers, as I was obsessed with the idea that I was going to college in September. I applied to the University of Idaho and was accepted. Eleanor Stewart was a student there, and through her, by correspondence, I obtained a job working for board and room with a Mrs. Frances Miller who had a house in Moscow, the site of the university.

In August, leaving the farm responsibilities to my younger brothers, I went to earn some money in the wheat harvest in the Palouse. The Palouse country in eastern Washington is a fabulously fertile farming area which is covered by quite steep hills—all of which are farmed on both the sides and tops. They are so steep that farming is all contour—that is, around the sides and on their tops—and the equipment has individually suspended wheels controlled by separate engines to lower the wheels on the downside of the hill sometimes by as much as ten feet! In those days, the combines cut and threshed the wheat, delivering the grain through a pipe, which was emptied into burlap sacks on a moving platform at the rear of the rig. These were then sewed to close the tops and kicked off the back of the machine to be picked up by crews following with trucks. This was all pretty specialized work, and at age sixteen with no prior experience with these harvesting techniques, I was considered (appropriately) a greenhorn. However, I was able to find short-term jobs as a tender for the horses and a pickup man for loading the sacks onto a truck—each of which weighed one hundred pounds. Also, I worked as header tender for the combine and as a truck driver. The work was heavy, and the pay was low! At the end of the period, I had been able to save thirty-five dollars.

The University of Idaho

And so with my money in my pocket and a Baby Ben alarm clock which the family gave me, I was off to college. The family drove the ninety miles to Moscow to deposit me at the home of Mrs. Miller with whom I had previously arranged to stay, working for board and room. However, when we arrived, she said that she had rented her house to the new dean of forestry, Dr. McCardle, and that she was going to California. I was given permission to stay in the basement of the house until I had made other arrangements—so long as I did not cook or prepare food there (fire hazard!) and if I worked on the yard for three to four hours a week.

My parents were very reluctant to leave me, but I was absolutely determined to go to school. We bid each other sad good-byes, and I was on my own.

Mrs. Miller's house, it turned out, was three miles from the university campus—down through the center of Moscow then up the hill to the university. Both sides of the valley were fairly long climbs! I had also prearranged a job to work for the university in what was called the Federal Emergency Relief Administration (FERA) at thirty-five cents per hour. This meant working on the college farm or on ditch-digging or other construction projects on the campus. I was allowed to earn up to twenty dollars a month, but the paycheck did not come until two weeks into the month following the one in which I earned the money.

On the morning following my arrival, I walked over to the university to register. The throng of students—not one of whom I knew—shocked me. My class of thirty in high school had not prepared me for this! I registered as a freshman in the school of education, signing up for English, psychology, history, educational theory, physical education, orchestra (trumpet), and ROTC (Reserve Officers Training Corps).

So after registering, I was sent to the bursar's desk. My bill was forty-nine dollars and fifty cents! Of course, I had thirty-five dollars in my pocket! When I got to his desk, he said, "Your bill is $49.50. Would you like to pay it all now, or part now and the rest later?"

I didn't know you could do that! But I thought quickly, *I must not let on how badly off I am, or someone is sure to want to send me home.* So I said, "Well, I will pay part now and the rest later."

"How much do you want to pay now?"

"Twenty-five dollars now, and the rest later." He accepted the payment and gave me a bill for the residual!

So I was in! I began attending classes, working on the college farm, working on Mrs. Miller's yard, walking back and forth from my basement room, going back to the college library at night because I had no money to buy my books, and eating in a restaurant in the town as I passed through.

I was very careful to make economical food choices. At first, I ate two meals a day, then one, then one meal every two days. When letters from my family came, I took off the stamps and put them on my replies, giving them no information about my increasingly dire straits.

My professor of psychology, Dr. Barton, the chief of the department, was very upset about people who were not serious students and kept making sarcastic remarks about those of us who did not study. Of course, I thought he was addressing me. Who could be studying less than I?

On a particular Saturday morning, I was scheduled to work on the college farm. I walked over there but just didn't have the strength to work. Every joint in my body ached, my eyes ached, and I had very little strength. So I went back to my room, but as I passed through the town, I stopped in a grocery store to see what most nutritious food I could get for the eighty cents I had in my pocket. I settled on a can of Ovaltine and a quart of milk. I lacked five cents for the deposit on the bottle, but I talked the owner into letting me have it. *If I must starve*, I thought, *I want to put it off as long as possible.*

I walked home, stirred all the Ovaltine I could into that quart of milk, drank it, and lay down on my couch. No soul in the world knew my problems. I was being set up for another miracle! I had no contact whatever with the McCardles since I came and went from a basement door. But after I had lain for about two hours, Mrs. McCardle came to the head of the stairs and called out, "Mr. Rodkey!"

"Yes."

"Have you been able to make other living arrangements?"

"No, unfortunately, not yet."

"Well, Dr. McCardle and I have been trying to find a young woman to live in and help with the meals and the children, but we have not been able to find anyone who is satisfactory. Would you be willing to help out until we can find someone?"

"Why yes, I would!"

"Good. Please come on up. I have some stairs that need to be washed down!"

From then on, my schedule was even more complex—three hours of housework each day added to my previous routine. By this time, we were coming to midterm evaluations. In the school of education, they did not give out grades, but we students filed by and looked at our grades on a card file in the dean's office. When I came by, I simply could not believe what I saw! The grades were all handwritten, and all were As except for a B in physical education. I knew that was wrong, and I thought, *How cruel that someone could write Fs that look so much like As!* Then the line of following students pushed me on. I fully believed I was failing until the end of the semester when I got a paper in my hand that said otherwise!

By the end of November, Dad and Mother realized that I had too much time required in work to support myself. They and Granddaddy took the lining boards from the granary on his farm and built a small cabin—seven feet wide, six feet tall, and twelve feet long. They loaded this on Uncle Royal's trailer and brought it down to me. The university allowed squatters with cabins to park them on university grounds adjacent to the power plant, and they supplied us with electricity and running water (from an outside spigot). So we parked my cabin there, with a folding cot, a monkey stove, a sack of apples, and a sack of potatoes for food!

This freed up some time. I was still working on the FERA program, but I was much nearer the college library. However, I was not feeling well. I saw a doctor at the University Health Service who diagnosed severe malnutrition and outlined a new (and very expensive) diet, which I should take. Of course, I did not have the money for this, so I went to one of the grocery stores in the town center, run by Mr. LePard. I told him my story and asked if he would be willing to extend me credit. He looked at me very thoughtfully, and then said, "I have had pretty good luck with university students. I will take a chance on you." So I bought the foods I needed, eventually, running up a bill of nearly $200, which I paid off completely by the end of the year. And I regained my health.

Studying in my cabin one night near the end of December, I heard a knock on the door. In came Gene Bouchard with a tale almost as bad as mine. He had one semester of study, was forced to drop out because of a lack of money, and was anxious to get back in to the second semester. He had applied to all the other cabin occupants, and no one would take him in. Would I? I considered very quickly. I certainly would not have had my opportunity except for the generosity of others. Surely, I had an obligation to pass it on. So I agreed.

Gene had less money than I, so I became the banker! We both registered for the second semester on partial payment. I bought the groceries from Mr. LePard on my credit account, and we both did our studying in the library. We finished the year in this way. I paid off Mr. LePard. Gene paid what he owed me, and then I bought all my books for the year and took them home to study during the summer. And then I had just $35 left over! Incredible but true! When I finally got my grades at year's end, I had a 3.92 grade point average out of a possible 4.0.

The interactions with Dr. Barton, my psychology professor in the first semester, were significant. I have mentioned his scorn for slothful students. At one point, he gave the class some sort of a combined intelligence-aptitude test, afterward inviting each student to come see him for vocational counseling. When I went to see him, he studied my scores very carefully, spending several silent minutes, then said, "Well, Rodkey, you scored in the ninety-eighth percentile on everything. So far as I can see, you can do anything you want—perhaps with the exception of music. And for all I know, you can do that too."

Several months after the semester ended, I happened to meet Dr. Barton on the steps of the administration building. He stopped and said, "Rodkey, did you ever think of going into medicine?"

"No, sir."

"Then think about it!"

And with that, he was gone, and I never saw him again!

Also, in retrospect, my consistent library study had another important benefit: I had access to the world's leading newspapers. I read the accounts of the rumbling unrest in Europe, including the rise of Hitler's Brownshirts and the plebiscite of Alsace-Lorraine. I became quite aware of the mass killings in Russia and of Japan's invasion of China with its accompanying atrocities. There were occasional student discussions of these issues as well. These stories and discussions conveyed a sense of dark foreboding!

Chapter 7

Summer of Change

With the ending of school, I was anxious to get home. But the winter had brought many changes there too. Dad, Granddaddy, and my brothers had begun logging on the top of a mountain south of Post Falls—another activity totally foreign to plains ranchers! Dad had bought the stumpage on an acreage of trees, which were required to be cut down and transported into the valley for firewood. This involved felling the trees, cutting off the limbs, sawing the trunk into sixteen-foot lengths, then transporting these logs by truck down the mountain to a level work lot where they could be cut into four-foot lengths and split for firewood.

As part of the preparation, Dad and his crew had prepared a log chute to carry the logs from the very top of the mountain about four hundred feet down to a staging area where we could bring in a truck. After a lightning slide into the staging area where they bounced around like toothpicks, logs were then loaded onto the truck and carried (in compound low gear) down the mountain to a vacant lot in Post Falls. There, we completed the breakdown into cord wood, then delivered and stacked the finished product for a delivered price of four dollars per cord—a pile measuring four by four by eight feet. From beginning to end, this was all handwork—there were no power tools then! As Granddaddy said, "Firewood warms you twice—once when you cut it and then when you burn it!" We were all warmed enough that summer! The work was both incredibly hard and dangerous.

After Labor Day, the younger children went back to school, and Dad and I were left to complete the logging work before I should go back to the university near the end of September. Then another miracle changed my life.

Dad and I were loading the truck with logs, and we had put on three fair-sized logs when we came to a twelve-foot-long red fir butt log measuring about four feet across at the large end and tapering sharply toward its tip. Such a log has to have its small end thrown far down the slope in order for it to be squared with the truck when it rolls up onto the bunkers. How far down is a matter of judgment. Dad and I disagreed. I knew the tip had to go much farther down than he thought, but he was very obstinate. We argued, and finally, in order to avoid coming to blows, I cut it loose as he wished. I had been right, and he had misjudged. Moreover, he got in the way of the log, and it ran directly over him, crushing him completely into the earth, so he was spread out looking like a squashed frog—his surface flat level with the surrounding earth. I dug him out of the ground. He was still alive, but barely. I laid him out carefully and began to prepare the truck to take him down—our only means of obtaining help.

If a logging truck does not have a large enough load for the logs to be chained securely to its bunkers, the loose logs will sheer through the cabin as one drives down a steep grade. Therefore, I had to get the three logs off. The bunkers were tilted in toward the bank in order to help avoid having logs roll over the outer side of the truck during loading; but if you roll the logs in against the bank to unload, they may push the truck over the outer shoulder of the road. Thus, I had to unload those three logs by rolling them up the outer side of the bunkers and throwing them off down the mountain. This required every ounce of my strength and more, but I did it. Then I tied Dad into the cabin of the truck (there was no door) and drove in compound-low gear down to Post Falls.

I drove to the doctor's office where, at his direction, we carried Dad and placed him onto a couch which was sitting on his outside porch. From time to time, the doctor emerged to look at him and then returned to his other work. Of course, I knew nothing about such matters then, but in retrospect, Dad was in deep shock, cold, and probably had injured his spleen and his liver and was bleeding from his pelvic venous plexus. Finally, at the end of the afternoon, the doctor said, "I can't do anything for him. Take him home, and if he is still alive in the morning, we will take him to Spokane for x-rays!"

He made it, and x-rays showed multiple pelvis fractures for which he was required to have six weeks of bed rest in a pelvic sling then three months on crutches.

Chapter 8

CHANGE IN DIRECTION

There was no alternative. I had to stay home to support the family and forego my return to the university. First, the remainder of the logs had to be cut, split, and transported to fulfill the Post Falls School's contract for two hundred cords of wood. Granddaddy Rodkey helped me as much as he could, and Lee, John, and George pitched in after school hours, and we got the job done. I rented my cabin to two recent graduates of Post Falls High School—Chuck Tiller and Wendell Satre. Wendell later became a renowned electrical engineer and, for many years, head of the Washington Water Power Company in Spokane.

We fulfilled the Post Falls School's contract!

In late October, our farm lease was expiring, and we learned of a farm for sale up in the mountains two miles southeast of Post Falls—the Gordon Farm. I took Dad and Mother to inspect the place, and they bought it for $1,200—on a mortgage, of course, but ours! I moved the family belongings with a team and wagon—the same Buster and Belle and wagon which Lee had driven over the hill to escape the auctioneer in Limon! (As we later learned, we were only the third owners of this land since its removal from Indian territory.)

The water supply for the house and livestock came from a spring about one-quarter mile up the mountain, but the pipes were so silted that only a trickle came through. I dug up the entire line, took apart each twenty-foot joint, swabbed out the silt, reassembled the pipe, and reburied the line. All the while I was working, I had water running in from the spring and flowing through reassembled line. Despite this, we had a fluke very cold mid-October night which resulted in freezing the entire segment of pipe through which water was flowing! We had no money to replace the pipe! Therefore, I took apart each joint, cut out the burst segments, and put threads and connectors on the remaining pipe to salvage all I could. In the end, I had to buy only forty feet of pipe, and the job was successfully completed!

There was a fertile ten-acre hay meadow on the farm, long neglected, which I plowed with the team, and walking plow—all (horses and plow) had been my companions on the immigrant car! Meantime, Mother, Dad, and the younger children made the house habitable, and the older children continued school in Post Falls.

Dad bought—for twenty-five dollars—a surplus racetrack barn down in the Spokane Valley directly across the Spokane River from our property. This was a building measuring twenty-two by sixty feet. I dismantled it and carried the boards to a sawhorse where Dad (still on crutches) knocked out the nails. I then loaded the boards onto the wagon, and we hauled them up the mountain to the farm and stacked them for later usage. These boards provided the lumber that eventually built two new chicken houses—each twenty-two by thirty feet!

Winter set in, and outside, work was no longer feasible. In Spokane, the US Army maintained Fort George Wright, which, at the time, was administering the Civilian Conservation Corps (CCC) Program. This program took indigent young men from the eastern cities to work in the western forests, cutting down brush and combating the blister rust fungus, which was destroying white pine forests. The army provided transport, housing, command, and medical care.

The hospital at Fort George Wright was very busy with various illnesses and accidents among these young men and was hiring civilian help. The commanding officer of the hospital, Col. Guy Guthrie, a graduate of the University of Vermont Medical School, had known my parents in their childhood home of Irving, Kansas, and he gave me a job as orderly in the hospital.

The hospital work was fascinating, and I quickly learned to do all the technical jobs in the place, including x-ray and laboratory work. The enlisted corpsmen were delighted to teach me so that they could have coverage for weekend passes!

Among other problems, we could not get a barber to cut the hair of our patients. So I bought clippers, barber scissors, and comb and became a barber—a trade which has since stood me in good stead!

One of the surgeons, Dr. John Wiltsie Epton, was a graduate of the Harvard Medical School. He said, "Grant, you may want to go to medical school. If you do, don't forget the eastern schools. They cost more, but they have better scholarships."

Spring of 1936 came and logging wages were higher than those of a hospital orderly! I quit Fort Wright and went to work in the woods for Mr. Frank Schintzel along with my brothers Lee and John. We spent the summer at this, and after Labor Day, Lee and John returned to high school. I still had two weeks to work before University of Idaho's classes began. However, as I was chopping limbs off a felled tree one afternoon, Dad came walking up the mountain with two young men representing Whitworth College in Spokane, Washington. Keith Murray and Loren Hatcher gave me a description of the college's advantages and urged me to switch schools. I was very reluctant to consider this, but finally, I realized that Dad and Mother seemed very anxious for me to make the change. So I picked up my ax and my lunch bucket, walked down the mountain with them, changed clothes, and they drove me off to Whitworth!

In 1987—fifty-one years later—at a meeting of the council on Medical Service of the American Medical Association (of which I was then chairman) in Seattle, I had a dinner party in my hotel for some twenty-five western Washington friends whom I had met at Whitworth—among them, Dr. Keith Murray, by that time emeritus professor of history at Western Washington College. Keith told me a story which completely shocked me: Col. Guy Guthrie of Fort George Wright had called Whitworth College and said, "You must not let this fellow get away from you." Thus, Keith's forest entrapment of me!

Chapter 9

WHITWORTH YEARS

Whitworth then had about two hundred students and a highly able, generous, motivated, but underpaid faculty. I began a premedical curriculum, compressing a four-year track into three years. For my biology major, my professor was Leslie R. Hedrick, PhD, from the University of Michigan, a brilliant researcher and teacher. For my chemistry minor, my professor was Mr. Benjamin Neustel, who had extensive experience in industrial chemistry. The class size seldom exceeded five to six individuals, so there was almost a tutorial experience throughout the entire course of premedical training. (Total student body enrolment of the school at the time was two hundred.)

The remainder of my college experience was very happy and productive as well, with more cultural courses, music, inspirational faculty, and many friends.

Brother Lee joined me in my second year at Whitworth. We brought up the cabin from Moscow, Idaho, for our living quarters, and we toiled at all the jobs in the place. I became the college barber as we scratched to pay our tuition. "Two evenings a week, twenty-five cents a cut, four/hour, for both students and faculty!"

This shack provided humble college housing for the Rodkey boys during their college years. Here, Grant and Lee move in a piece of furniture when the shack was located across the street from the main entrance to Whitworth College.

The role of music (mainly vocal) during our Whitworth years was dominant. Ms. Winifred McNair Hopkins was head of that section, which included choruses, small groups, and individual performance instruction. Lee and I sang in the chorus for Gilbert and Sullivan operettas (I was once Nanki-Poo in the *Mikado*!), Christmas performances of Handel's *Messiah*, spring presentations of John Stainer's *Crucifixion*, and many others. I was also one of the tenors in the Whitworth male quartet. Many of these performances were taken on the road to other regional communities.

An interesting sidelight to our road trips was our opportunity to drive through the Columbia River valley and watch the progress of construction of the Grand Coulee Dam. Begun in 1933 and completed in 1942, this was one of the largest construction projects in the world, and it is the largest electric power generating facility in the United States. At the time of our visits, there was a new construction town below the base of the dam (Mason City) built to house approximately 2,500 laborers. The dam was planned to be 550 feet in height, nearly a mile long, and was to create a lake backing up one hundred miles and crossing the US-Canadian border. Timber workers were cutting a swath of trees on either side of the Columbia Valley along the planned surface level of the projected lake—twenty feet above and twenty feet below the anticipated waterline. It was an eerie sight to drive along the highway along the Columbia River and look high on the mountains on either side and see those "bald" stripes with the realization that all between them was to be underwater! Huge quantities of timber were harvested in this endeavor!

The Grand Coulee Dam also provides water to irrigate nearly three quarter million acres in the Grand Coulee—an arid area scooped out by waters released during glacial melting at the end of the last ice age. The magnitude of this entire project is nearly incomprehensible! The chief advocate for its construction was memorialized by naming that great lake the "Franklin Delano Roosevelt Lake"—a significant accomplishment of our fated century!

During the Christmas vacation of 1938, in the Spokane Public Library, I came by chance upon a copy of the *Harvard University Catalog*—a publication previously unknown to me. Picking it up and reviewing the medical school section, I found that John Epton had been correct; there were some excellent scholarships! So I wrote to the dean of admissions, Dr. Worth Hale, and requested an application form. He sent the form but indicated that they could not consider it until I had taken the medical aptitude test—which I had not done, having no hope that I would be able to go on to medical school without some hiatus years to earn and save some money.

However, Dean Francis Tiley Hardwick said, "That's no problem. You can take the test next Saturday in my office, and I will monitor you."

And so we did!

Within two weeks, Whitworth College received from Harvard a request for a copy of its catalog. In another fortnight, I received a letter from Harvard saying, "You have been accepted into the medical school entering class in September. Please send $50.00 to assure that you will attend." Of course, I did not have the money, but my professors said, "Grant, here is the fifty dollars. You must go!" Dean Hardwick introduced me to Mr. Smith, owner of Smith's Funeral Home, who generously gave me a loan to get started, and thus it began!

(In 1998, at our 55th Harvard Medical School Class Reunion, Dr. Sidney Ellery, by then a retired surgeon from Long Beach, California, said, "Grant, there is something which I should have told you many years ago. During our first year in medical school, when you were waiting on tables in the Vanderbilt Hall dining room, I had lunch with Dean Worth Hale at your table. He said to me, 'So we get a letter from this kid in the west who went to a college we never heard of, and he comes up with the highest score in the nation this year on the medical aptitude test!'" God rest his soul. Dr. Sidney Ellery died in the spring of 1999.)

After Whitworth graduation in May 1939, the college offered me a summer job of contacting students who had made inquiries about the school. They gave me a new, dark-blue 1939 Plymouth sedan, a thick pack of names and addresses to contact, a salary, an expense account, and a *territory*: Washington State, Northern Oregon, Northern Idaho, and Montana! This was more than a time zone in latitude, and I was up to it! This job lasted for two summers. During each one of which, I logged more than fifteen thousand miles over that immense, intensely beautiful patch of the world! I traveled to nearly every town and hamlet in the region, and my mind still stores the images of their beauty. In the course of events, I met hundreds of talented, ambitious, and productive young people—many of whose friendship I cherish to this day.

Among these struggling students was Carl Blanford of Post Falls, Idaho. Carl was the eldest among a family of six children. He had graduated from the Post Falls High School in May 1939. During his high school years, he had worked as an apprentice for Mr. Henry Wetherell—owner, publisher, and editor of the weekly *Post Falls Tribune*. As was the case with most of us, the Blanfords had to scrape pretty hard to make ends meet, and they were in no position to send Carl to college.

I went to the Whitworth authorities with the suggestion that they give a scholarship to Carl if he would become the college printer and save all the money Whitworth had to expend on the costs of its publications. Then I went to Mr. Wetherell and asked him for his suggestions as to how to proceed. He said, "In the basement, I have an old hand-operated press that is in working condition, and I will donate it to the college." Thus Carl got his scholarship, Whitworth got its print shop, and the world got a great benefactor! Carl was a brilliant student, graduated with honors, was accepted at Princeton Theological Seminary, and became a missionary in Southern China until the Chinese communists drove him out. He moved to Chiang Mai, Thailand, where he served until his retirement then worked for several years as assistant pastor in First Presbyterian Church in Seattle, finally moving to the Presbyterian retirement home in Duarte, California. He volunteered as an assistant pastor in a Chinese church in Duarte until he died in 2012. He's a great and good friend and benefactor to the world! And also—with his wife—was one of my hotel dinner guests in Seattle during the 1988 meeting of the council on Medical Service of the American Medical Association!

Chapter 10

MEDICAL SCHOOL YEARS

In September 1939, I boarded a Greyhound bus in Spokane and rode for five days and five nights (with only rest stops) to Boston—a new world culturally and linguistically! I lived in a sixth floor walk-up room in Vanderbilt Hall (the medical school dormitory), waited on tables in the dining room, and began my studies—all pretty rigorous. Our class was made up of 125 individuals from across the world, with whom I developed lifelong friendships not unlike those comradeships that develop in the heat of battle! In particular, two became essentially new brothers: Jesse Thompson from Brownsville, Texas, and Harvey Frazier from Spokane, Washington. Jesse was a Rhodes scholar, trained later as a neurosurgeon, and became one of the world's most distinguished vascular surgeons. He was the father of carotid artery surgery for the prevention of strokes. He lived and worked in Dallas until his death in 2004.

Harvey Frazier came from and returned to Spokane, Washington, where he became a practicing obstetrician and assistant professor of obstetrics at the University of Washington, which sent some of its obstetrical residents (from Seattle) to Spokane for his teaching. Harvey was the originator of the underpublicized *Prayer Prescriptions* which addressed the core emotional and humanistic needs of his patients. Color blind in the extreme, Harvey became a powerful painter, cutting to the core of emotions and self. In 1992, he wrote, "Since I have been doing this (painting) I have been enabled to see a new picture of medicine and the ministry of healing! The darkness of despair and depression is broken asunder. Beyond we see the beauty of eternal life face-to-face. My patients have taught me about the healing power of God's love! My patients have taught me about prayer and healing. My patients have taught me about scripture prescriptions." The Boss called Harvey home in the spring of 2000.

And so these remain my newly adopted brothers into eternity!

Medical school was tough. I tremendously enjoyed my professors, the challenge of new learning, and eventually, the contact with patients. I remained economically challenged. In my second year, I was fortunate to join a group of eight students who were hired as student house officers at the New England Deaconess Hospital. In essence, this meant doing the history and physical examination of patients admitted to the Lahey and Joslin Clinic's inpatient hospital services and providing emergency laboratory services for the hospital, including blood and urine analysis, typing and cross matching, transfusion, and parenteral infusion services. Our pay was board, room, and laundry. In addition, I worked as night watchman at the Harvard Medical School—two nights per week. These jobs continued until my graduation. I was also blessed to receive some of those good scholarships described by Dr. John Wiltsie Epton—thanks to the generosity of Harvard Medical School.

During the second year at Harvard, my life took another fateful turn. For totally inexplicable reasons, and with no prior contact with anyone who had experience in the Orient, I had entered medical school with the idea of working in medical education in China. In 1940, I was introduced by Harvey Frazier and another of our classmates, Roger Morrison, to Dorothea Smith, Wellesley '41. Dorothea was the daughter of Dr. C. Stanley and Dorothy Smith, distinguished Presbyterian missionaries of Nanking, China. We fell in love. In 1941–42, she took her master's degree in Christian education from Columbia University. At the time of her graduation, her parents were imprisoned by the Japanese in a concentration camp in Shanghai. She had no close relatives nearby in America, no job, and no money. So we were married in June 1942! We rented an apartment adjacent to the Deaconess Hospital, she found a job as claims adjuster with Liberty Mutual Insurance Company, and I continued my jobs until graduation.

In the early years, Dorothea worked as a social worker to help support me. Later, we had two children—Cheryl Anne and John Mark—in whose upbringing she played the major role. After they left home, she continued as a volunteer leader in the Staff Wives of the Massachusetts General Hospital, in Amigos de las Américas, in the Massachusetts Medical Alliance, and in the Plymouth Congregational Church in Belmont, Massachusetts—all of which institutions were greatly benefited by her competence and devotion.

In the summer of 1942, I was hired by Dr. Henry K. Beecher, chief of anesthesia, Massachusetts General Hospital, to give anesthesia to patients—at that time still, mainly, drop ether. This was a remarkable opportunity to learn, to meet the surgeons of the MGH, and to make a little money! When it came to the internship match of the spring of 1943—thanks to Dr. Beecher in part, I believe, I was selected as one of eight surgical interns chosen as the annual quota by the Massachusetts General Hospital. (There will be more comments regarding Dr. Beecher later in this dissertation.)

Because I had decided to apply for a surgical residency, I took my fourth year rotation in surgery early in the academic year of 1942–43. Thus, it happened that I was serving at the Massachusetts General Hospital at the time of the Cocoanut Grove Fire on November 28, 1942. This was a horrendous nightclub fire which killed outright 492 patrons and sent scores more to all the hospitals in Boston—with burns, inhalation injuries, trampling injuries, and emotional shock. All major hospitals in Boston, including the Massachusetts General, Boston City Hospital, Peter Bent Brigham, and Beth Israel were overwhelmed by the dead and the dying. At the MGH, the brick corridor area (about fifty by one hundred feet) was ringed by corpses laid out on the floor awaiting identification. The emergency ward was totally inadequate to accommodate the crush, so the entire sixth and twelfth floors of the White Building were converted to burn units, with the smaller twelfth floor being concentrated on respiratory tract burns. This event, occurring just one year following the Pearl Harbor attack and the entry of the United States into World War II, found most Boston surgeons in the prime of their careers away on military service in the European or Pacific theaters of war. Thus, extraordinary responsibilities were thrust upon the remaining attending surgical staffs and upon surgical residents and medical students.

Fortunately, Boston was, to a certain extent, prepared for the emergency. One of the grave injury problems identified in the Pearl Harbor attack had been extensive burns. Dr. Oliver Cope, of the MGH, had received, thereafter, a federal grant to study the treatment of burns. His work had concentrated upon early treatment with bandages of bland petrolatum ointment which proved to be far superior to earlier concepts of burn wound management. Also, this was at the time of initial clinical trials of penicillin—the first widely effective antibiotic. Lessons learned in fluid resuscitation, respiratory support, blood replacement, infection control, plastic surgery repair, nutrition, and rehabilitation benefited the nation and the world, but the cost was horrific.

My fourth year rotation in medicine was at the Peter Bent Brigham Hospital in January through February 1943. As it happened, the resident in charge of my assigned ward was Dr. Curtis Prout, Harvard Medical School Class of 1941. The ward was full and the demands on medical residents were proportionally greater than normal because of conscription of their attending physicians into military service. Thus, it happened that medical students were also given some unusual assignments. In particular, we received a young man (aged nineteen) from Maine who had been cyanotic from birth and who had severe limitations in exercise tolerance. We discovered that the condition was caused by congenital methemoglobinemia—a rare condition caused by conversion of hemoglobin to its oxidized form which seriously inhibits oxygen transfer from red blood cells to tissues. No treatment was known. Dr. Prout assigned his care to me: "You plan what to do for him."

Quick review revealed that there were no clues in the medical literature. This was, innately, a chemical problem, so I reverted to my training in organic chemistry with Dr. Benjamin Neustel at Whitworth College and with Dr. A. Baird Hastings, my professor in biochemistry at Harvard Medical School. There germinated in my thinking the idea that a nontoxic catalytic reducing agent might be useful. One such agent, potentially, might be vitamin C. Emboldened by the idea, after consulting Dr. Prout, I injected intravenously a dose of vitamin C into our patient. There was no immediate apparent effect (although later studies proved that vitamin C [ascorbic acid] does, indeed, reduce methemoglobin to hemoglobin but the rate was too slow to be visually apparent). I then considered methylene blue—an organic compound capable of accepting an oxygen molecule and converting to a leuco form. This substance had been used safely as an adjunct to study of the kidneys. So I tried an intravenous injection of methylene blue. Bingo! Within minutes, the patient began to change color from purple to bright red! The patient himself was entranced! He ran to the bathroom to study himself in the mirror, and he could not be induced to leave for many hours! This was the first time he had ever seen himself pink like the rest of humanity!

At Dr. Curtis Prout's direction, we presented the patient at the next week's Medical Grand Rounds. Dr. Prout gave the medical history and presented the purple-black patient. I then took the patient behind a cloth screen and injected methylene blue. A few minutes later, we took aside the screen and presented the transformation! It was a sensation! There was pandemonium in the gallery! Dr. Samuel Levine, chief of the cardiology service, drew Curt Prout aside after the meeting, saying, "This was the best Grand Rounds that I have ever seen at the Brigham!"

A few weeks later, I was invited to present the case at the second Undergraduate Assembly (now the Annual Soma Weiss Assembly). I remember seeing Dr. Walter B. Cannon (physiology) and Dr. A. Baird Hastings (biochemistry) sitting in front row seats and beaming broadly! They were delighted at the clinical application of their teachings!

Perhaps the most amazing aspect of this whole story is the fact that methylene blue is still, to this day, the prime treatment for methemoglobinemia! Although we had no access to the information in 1943, as is the case with so many innovative ideas, the same discovery was made independently in several other clinics in the world at approximately the same time.

Chapter 11

RESIDENCY YEARS—PHASE ONE

Because of the entry of the United States into World War II after Pearl Harbor on December 7, 1941, shortly afterward, I had been drafted into the army as a second lieutenant—with all my classmates. Vacations were abolished. Therefore, we graduated on March 31, 1943, and went to work as interns on April 1. The training was rigorous. Our schedule was alternate nights on and off, but on both on/off shifts, we were required to sleep in the hospital. Pay was board, room, uniforms, and laundry. Military duty had stripped many of the surgical staff from the hospital, so there was a very heavy workload for those left behind. This gave early, steeply graduated responsibility to the residents, and I rapidly advanced in the ranks. During my third ninth-month rotation, I became chief resident on the orthopedic service under the direction of the great Dr. Marius N. Smith-Petersen—the pioneer of hip replacement surgery. My new "brother" Jesse Thompson had a similar trajectory and was, during the same period, the chief resident in neurosurgery under Dr. Jason Mixter.

Also, I must introduce Dr. Chiu-an Wang. Chiu-an had been born in Canton, China, the son of a Chinese Lutheran bishop, and the eldest of a family of ten. He had received a master's degree in parasitology from Lingnan University in Canton and had spent two years in Peking Union Medical College. His father's sister, the first Chinese woman trained in Western medicine (and working in Canton), decided that Chiu-an should go to Harvard Medical School. Thus, he arrived as a member of the class immediately behind mine. He also received an internship appointment at the Massachusetts General Hospital and so began his rotations exactly one step behind me. As a Chinese citizen, he was exempt from American military duty. We worked together regularly and became extremely close friends—another adopted brother! During this interval, he met and married Alice Sze. Alice had been born in Washington, DC, the daughter of the Chinese ambassador to America (1918–1950). Alice was a graduate of Wellesley College and was working as a technician in the MGH blood bank at the time she and Chiu-an met.

During the eighteen months from January 1944–June 1945, Chiu-an Wang's rotations were just behind and parallel to mine, so I became his immediate supervisor. We found ourselves, citizens from opposite sides of the globe, to be remarkably similar in attitudes, talents, and motivations and spent many happy hours working side by side. But we were separated by calls to active army service for me and for Jesse Thompson in July 1945.

Chapter 12

MILITARY SERVICE

Our first six weeks of military training was an assignment to Carlisle Barracks, Carlisle, Pennsylvania. Here, we learned the rudiments of military ethics, courtesy, marching, commands, uniforms, attitudes, and behavior. Jesse and I were not in the same platoon, but we saw each other frequently. Each of us had brought his wife, so we lived in off-base housing. As it happened, I owned a second-hand 1933 Ford V-8 cabriolet, so transportation was not a problem—except that gasoline was rationed (along with sugar, butter, meat) so that we could buy only ten gallons per week. Compared to the stringency of our work schedule at the hospital, this duty was much easier!

V-E Day (Victory in Europe Day) had been celebrated on May 8, 1945, prior to our entry into active duty status. While we were at Carlisle Barracks, the atomic bombings of Hiroshima and Nagasaki occurred on August 6 and August 9. On August 9, the Soviet Union declared war on Japan. On August 14, 1945, President Truman announced the formal surrender of Japan under terms of the Potsdam Agreement, and a bedlam of celebration erupted! Every store, restaurant, gas station, and all other public facilities shut down for twenty-four hours! All of us had known that we were headed for the Pacific and mortal combat with Japan, and we were deeply relieved by the thought of a more hopeful future. I received orders to proceed to Camp Shelby, Hattiesburg, Mississippi, where I was assigned to the separation center as an orthopedic consultant. Jesse had a more interesting assignment. He was assigned to a neurosurgical hospital in the Greenbrier Resort in West Virginia. Because of our abbreviated surgical chief residencies, we were given US Army specialty ratings—Jesse in neurosurgery and I in orthopedic surgery.

My orders to proceed to Camp Shelby were accompanied by a waiver on gas rationing to make the trip possible. Dorothea and I headed off to Dixie!

Parenthetically, she had planned to get her driver's license while we were at Carlisle and had carefully read the manuals, studied the laws, and practiced with our Ford. When she took the licensing examination, during the driving test with the state trooper, she set off confidently, driving on the left side of the road! This was a reversion to her China days where left hand drive was all she had known! (China changed from left to right hand drive on July 1, 1945.) Needless to say, I was the driver for the trip to Mississippi!

This proved to be a significant journey: through Maryland, Virginia, North and South Carolina, Northern Georgia, across Alabama, and down to Southeastern Mississippi—Appalachia. The scenery was beautiful; farming methods and crops were new (tobacco, cotton, rice, peanuts, pecans, peaches); and language, dialect, and some customs were new to us. But our brisk travel schedule precluded more than superficial observation. When we arrived in Hattiesburg, it was necessary to find civilian housing—no quarters were available in Camp Shelby. Available housing was scarce and consisted mainly of an individual room in a private home with a going price of ten dollars per week. After searching for nearly half a day, we came across a home with a sign in the window advertising such a room. I walked up to the house which had a screened porch with a locked door and an intervening space of about ten feet before the main door of the house. I knocked firmly, waited a few minutes, and knocked again. The lock on the inner door was opened from within the house, and a woman appeared. She looked me over very carefully then asked, "Are you from New York?"

Taken aback, I replied, "No, I am from Boston."

Whereupon, the woman unlocked the screen door and invited me in! Fortunately, this was not a characteristic reaction of the citizens of Hattiesburg. After a few weeks, Dorothea and I moved to a room beneath the college football stadium. These were nice quarters despite the bedlam of noise during Friday night football games! We attended a local church, sang in their choir, and made many good friends among the townspeople.

Camp Shelby encompasses approximately two hundred square miles. At the time I was there, it was used primarily as a training center for troops destined for overseas duty. (Parenthetically, there was a current saying that anyone who volunteered for overseas duty from Camp Shelby was simply yellow [a coward]!) However, following the end of hostilities with Japan, Camp Shelby had been converted to a separation center for army and air force troops being demobilized. I was assigned as an orthopedic consultant to assist in evaluating injuries or illnesses contracted during military duty. Days were full, as there was a huge stream of soldiers being demobilized. The work was heavy, but not stimulating. I learned later that a friend of mine from Post Falls went through the line, but I saw only his feet!

But times were changing. In early December, I had a telephone call from Jessie Thompson at the Greenbrier Hotel: "How would you like to be stationed up here as the orthopedic surgical consultant to this neurosurgical center?" Straight from heaven! "Well, the commander up here has specifically requested you to be assigned to that job. Begin packing your bags!" I hung up the phone and began packing—furiously!

Within two hours, the telephone rang again. "Rodkey?"

"Yes."

"This is the aide-de-camp to the commander at Camp Shelby. If you want any leave before being shipped overseas, be at my office at 8:00 AM tomorrow!"

"Yes, sir!" What had happened? Am I being shipped out because of my requested transfer?

On the next morning, I arrived at the office of the commanding officer at 8:00 AM. I received orders to take my wife by automobile to Boston, thereafter to travel by rail to Camp Pendleton, California, and three days thereafter to proceed by rail to Fort Lewis, Washington. In those days, military secrecy was still in force, so my orders were confidential.

We were on the road again! Essentially, we traced northward the same route which we knew and headed northeast through Washington, DC; Philadelphia; New York; New Haven; and Boston. However, this was winter, and there was considerable snow in the Appalachians. I tried to buy tire chains, but because of the wartime metal shortage, none were available. The trip was something of a struggle, but we managed it safely.

There was an unexpected bonus! Dorothea's father and mother had been on sabbatical leave spending the year at Yale. Her father had returned to China a few weeks earlier, traveling via Lisbon, Rome, and northern India to enter west China. Her mother was still in their sabbatical apartment in New Haven, and we were able to spend a night with her. In Boston, we rejoined my brother Lee, his wife, Marjorie, and their infant son, David. I left Dorothea and the car with them and headed west.

That transcontinental trip was an unforgettable experience! Current readers will not remember that during World War II, *all American citizens were involved.* All able-bodied men were in military service, and those left behind—including women—were involved in the industrial effort to support the war.

(Years later, in examining a patient with abdominal problems, I noticed a circular scar on her abdomen just above the umbilicus.

"Were you a welder in World War II?" I asked.

"Why, yes! How do you know?"

The scar was the reminder of a hot drop of metal which had splattered onto her.)

Throughout our entire journey, our passenger train was frequently placed on a siding to allow freight trains to pass through on priority schedules. These were great, long trains tightly loaded with trucks, tractors, tanks, many other types of heavy construction and earth-moving equipment, coal, ore, oil, corn, wheat, barley, and many, many other heavy freight units. Never before, nor since, have I had the overwhelming sense of America's might—nor of our unity as a nation—which was given to me on that trip.

Camp Pendleton was a brief blur. My immunization record was checked, and I was given every immunization known! I was incorporated into a group of twenty-eight medical officers—all first lieutenants like me—and was moved onto a rail shipment to Fort Lewis in Seattle. Seattle in winter is cool and wet! There was constant rain and mud everywhere! Not much remained to be done for us except to check out personal equipment, which included insect spray and mosquito nets. In spite of secret orders, it did not take a genius to understand that we were headed to the Orient. I called Dorothea and gave her a coded message: "I think that I am likely to see your father before you do!"

And then we were boarding a Liberty Ship which had been converted to a troop transport. We had about eight hundred troops with military police and their officers and our straggling crew of doctors. Even at the dock, the floors seemed a little unsteady. Before we were out of the harbor, I was seasick, but this subsided in one day. One of the medical officers—actually attached to a battalion of military police—whom I had known as a surgical resident at the Peter Bent Brigham Hospital, was really stricken by this "disease." He was confined to his bunk for the entire trip. Two years later, I met him in Shanghai. He looked terrible.

"Charlie," I said. "What's the matter?"

"I'm going home."

"Why should you be so sad?"

"You don't understand. My battalion has been assigned to go to Manchuria and pick up a shipload of Germans imprisoned up there, take them south around India, through the Suez Canal to Germany. Then we have to sail to New York!"

I know he made it. Years later, I saw his name attached to an air force operation in North Africa!

On the first day out of port, we were given instructional booklets in the Chinese language and began daily instructional classes to teach us a few essential phrases. As it turns out, the Chinese and English languages have *no* common bridges such as we are accustomed to with Latin, Italian, French, Spanish, or German. Bobbing around in our Liberty Ship, half seasick, homesick, and apprehensive about a totally foreign future and facing the task of comprehending and remembering the most difficult language in the world were not meant for morale-building! Nor were they a formula for success! We all struggled but emerged from our voyage with no useful understanding of the Chinese language—written or spoken!

After twenty-one days of floating around, bouncing through two major storms, and watching the graceful albatross skimming the waves, we finally docked in Shanghai on the bank of the Whangpoo River. Debarking was a weird experience. It really looked like the same earth we had left in Seattle, but nothing else looked the same! The streets were pandemonium! Pedestrians were carrying bamboo poles across their shoulders with suspended loads at either end; push-carts, pull-carts, rickshaws with passengers, pigs being carried in enclosed two-wheeled carts, bicycles beyond count but no automobiles; and pedestrians without number in strange garb—mainly padded pants and jackets (since it was quite cold) or long padded gowns—all hurrying along the way!

As it happened, our dock was very near our destination: the Broadway Mansions Hotel located at the juncture of Soochow Creek and the Whangpoo River. The hotel was a reasonably modern structure with a dining room on the third floor, the Shanghai Station Hospital on the fifth floor, and billets for the troops on the upper floors. All the upper floors were *frigid* since the Japanese had removed all the pipes and radiators for metal salvage/conversion to armament. Spring was not imminent!

Following our sign-in and assignment of quarters, I had the rest of the day off. Through Dorothea and her parents, I knew that there was a Presbyterian mission compound (settlement, enclosure) in the French Quarter, and I thought this was located on Nanking Road. This road began at the Bund and extended for several miles in a generally southeasterly direction. The Bund was the main wharf area and commercial center of Shanghai and was located within a few blocks downstream on the Whangpoo River from our hotel. So I decided to see if I could find this island of English culture!

The walk down the Bund was no less confusing. Thousands of people from whom I could not understand a single word and signs on the street were all (except for one) totally indecipherable. The only sign in English was unforgettable: Guaranteed Accurate to Cut Prepuce! But within a few blocks, I came upon Nanking Road, which had both English and Chinese signage! And remarkably, there was a streetcar line with a trolley sitting on the track, ready for takeoff. With some trepidation, I boarded through the front door. I had no Chinese currency, no knowledge of the fare, and no way to communicate with the conductor/operator. After a few seconds of totally incomprehensible conversation, the man generously waved me to the back of the car! As I started down the aisle, I looked up and saw my father-in-law, Dr. C. Stanley Smith, standing in the center of the car and holding on to a central post! He had come from New York City, through Lisbon, India, over the hump of the Himalayas, through western China, and—with the recent departure of the Japanese—to Shanghai. He arrived a few days before me! This was, of course, another miracle and an overwhelming imperative to study the Chinese language!

Before assignment of my duties, I was required to be interviewed by the commanding medical officer in Shanghai. This turned out to be Col. George Ellis Armstrong, Medical Corps, United States Army (later major general and surgeon general of the army). Colonel Armstrong himself was an orthopedic surgeon who had some of his training under Dr. Marius Smith-Petersen! He obviously knew a great deal about me—including the fact that I was married to Dorothea and that her parents would likely be returning to Nanking! Therefore, he told me I would be assigned to the Shanghai Station Hospital for six months, and that my permanent assignment would be the Nanking Station Hospital. This was my single meeting with General Armstrong!

Life at the Shanghai Station Hospital was busy and constantly instructive. We were the most complex military hospital in China, and our evacuation hospital was the (converted) Saint Luke's Hospital in Tokyo. We received injured/ill military personnel (army, navy, marines, air force) from China and Southeast Asia and treated them locally or triaged them out by air transport to Tokyo depending upon the severity of the problems. As the orthopedic surgeon, I had many—some very serious—injured patients. Several of my fellow troop transport passengers were also in the same category as I- recognized as medical specialists. These included the specialties of internal medicine, dermatology, pathology, radiology, anesthesiology, and neurosurgery.

As we became more adjusted to our surroundings and found some interpreters to help us, we began to explore the neighborhood. These expeditions included shopping for mementos and art items which included many Japanese items. The Japanese occupants of China had been assembled in Shanghai, then shipped home, the last having shipped out in December 1945. Most of their belongings had to be left behind and wound up in street bazaars—merely a passing asterisk in the drama of human affairs!

From the windows of the Broadway Mansions, we looked down and learned more of the culture of this new/old society. The people were tirelessly working—all with heavy tasks. There were few mechanical aids, rickshaws and carts excepted. However, they always seemed to be good-natured and friendly whenever we were in contact. A fascinating contrast to our past experience was the life-cycle existence in the sampans on the Whangpoo River. People eat, work, sleep, fish, and carry out all the functions of life—all on little boats with a small sail and a long oar/rudder held by a boatman standing at the rear of the craft.

On one occasion, looking down from above, I saw a huge crowd gathered in front of the hotel around a sampan tied up at the dock. There appeared to be two people aboard—a woman seated at the front (and apparently weeping) and a man standing motionless at the rear of the craft, holding a bamboo pole down into the water. The crowd was intent, totally motionless, and silent. What was going on? I stood, watching this suspended drama for many minutes, speculating on its meaning. Finally, the boatman pulled his pole out of the water; tied at its tip was an egg-shaped, pink mass which I suddenly realized was a baby! Who knows the story? Probably poverty—no money to feed another mouth; perhaps a girl—not so desirable or so economically effective as a boy. A shattering experience!

On a Sunday afternoon in early April 1946, a group of our medical officers checked a jeep out of the motor pool and set off on a sightseeing tour of a perimeter road on the outskirts of Shanghai. Roads were not well maintained. At a certain point, we hit a deep pothole, and the jeep was flung into the air but not overturned. I was sitting in the backseat and—fortunately—holding on to surrounding supports firmly so that I remained in the vehicle. However, I had a severe whiplash injury to my neck and immediately felt numb from my neck downward—although I still had motor function. Although the numbness quickly subsided, I knew this meant some degree of spinal cord injury plus/minus cervical spine injury.

We returned to the hospital where x-rays were taken, and I was placed in a plaster-of-Paris body jacket from my jaw to my hips. I was never given a report of the x-rays, but considering the outcome, I suspect they did not show fractures. However, I was to be immobilized for six weeks!

This dreary interlude was interrupted one day by the unannounced arrival of two visitors: Dean Sidney Burwell of the Harvard Medical School and Dr. Alan Gregg, Director of the Rockefeller Institute! They had been sent to China by the Institute to evaluate how the library of the Peking Union Medical School had survived the Japanese occupation. They had completed their survey and were heading home and, in Shanghai, had heard of my mishap! A generous gesture by two giants! They had found and reported that the library had come through the occupation 94 percent intact. Later (in 1947), I myself visited that library and was shocked to find that it was a much larger collection than the Harvard Medical School Library had been during my years there. This was subsequently changed at Harvard by conversion to the Countway Library of Medicine, which will be discussed more completely at a later point in this story.

In June 1946, in compliance with General Armstrong's orders, I was transferred to the Nanking Station Hospital. This was a period when Gen. George Catlett Marshall was serving as ambassador to China, and we had a military advisory group to Generalissimo Chiang Kai-Shek. The Nanking Station Hospital serviced this official American community (approximately two thousand personnel), as well as two thousand Chinese employees. The hospital was modest, consisting of one floor of a converted hotel wing and two Quonset huts. The operating room was located in the outer extremity of the hotel wing. The remaining rooms and one of the Quonset huts were wards, the other Quonset hut contained the pharmacy and the laboratory. We had one army nurse, an air force anesthesiologist, and an additional group of five general medical officers and one radiologist. The commanding officer of the hospital was air force Col.

Merrill Moore, a psychiatrist from Quincy, Massachusetts, who had a longtime commitment to writing a sonnet each day and was a dedicated photographer. In fact, he carried a Leica 35 mm camera suspended from his neck across his belly on which he kept snapping exposures (without focusing) wherever he walked! He was on detached duty to serve with General Marshall, and he lived in the embassy quarters. I do not remember ever seeing him in the hospital! But he was, in fact, a gentleman of great charm! The table of organization for the whole MAG Group was bizarre: one hundred twenty-seven full colonels and two second lieutenants! In fact, one GI enlisted man (an MIT student in real life) who was in charge of the local army radio station nearly got himself court-martialed for insubordination by playing during the breakfast hour, "There Ain't Nobody Here but Us Chickens"! ("Chicken" is a slang term for a "bird", or full, colonel.)

I, as the only surgeon, became chief of general surgery, chief of orthopedic surgery, chief of urology, and chief of obstetrics! The latter was very odd. In medical school, I had obstetrical training under Dr. Fritz Irving, chief at the Boston Lying-In Hospital. I had delivered ten babies in the community outreach program, and I had assisted at three Cesarean sections. At the MGH, I had delivered several babies who arrived before their obstetricians could get to the hospital and had assisted at a few Cesarean sections.

For some weird reason, Dr. Irving's textbook, *Safe Deliverance*, was the only medical book I had put in my gear when I shipped overseas!

One of my chief functions was to care for the Chinese employees. Many of them were highly trained (such as pharmacists) but most were cooks, cleaners, vehicle drivers, and similar service personnel. But they were all highly intelligent and friendly, and I learned a great deal from them. Regularly, I had a session with a Chinese tutor at 5:00 AM because I could not bear being unable to communicate with people around me. I was becoming passably competent in conversational Mandarin.

Rapidly, I learned a good deal of Chinese medical terminology as well. Another remarkable observation came to me as a surgeon: Chinese bodies are "tougher" than those of Caucasians. That is, their fibrous tissue is tougher and more tightly put together than we are. This observation has been sustained during many subsequent years of operations on both occidentals and orientals. I have no idea why this is so—except for speculation that hundreds of centuries of tough physical hardship may have favored this genetic selection. It is of great interest that current genetic research may be providing support for my supposition of this genetic modification (Cell. 2013 Feb 14;152 (4); 691–702).

Many of the military advisory group (MAG) officers had their wives in Nanking. This meant brisk business for the obstetrical service! During this exposure, I met many wonderful families and had some memorable experiences. Of these, one stands out as being unforgettable: a Chinese-American citizen born in West Virginia was stationed in Nanking as a captain, infantry, US Army, and was detailed as aide-de-camp to General Marshall. (I shall deliberately delete his name.) Earlier, while stationed in west China, he had met and married a Chinese woman (whose English name was Anita) of great beauty and charm. However, she was as slender as a toothpick! This lady became pregnant, and I was her obstetrician. It was a worrisome prospect. Would she be able to deliver a baby? Four weeks before term, I found the baby positioned in the uterus with head up and butt down. It's called a breech presentation—a bad deal since babies are supposed to come out headfirst. I performed what is called "an external version;" that is, I turned the baby around with its head headed south. At the next visit (and all subsequent visits), the same circumstance and treatment were repeated. I became convinced that the mother's pelvic brim was too small to permit passage of the baby's head.

Anita went into labor!

No progress. Considerable pain all day, all night. The next morning, same status. The accepted obstetrical teaching at that time (and probably still) was that if there was no progress during labor for twenty-four hours, a caesarean section should be performed. This posed a serious problem, a delicate political balance, and a critical professional issue. I had never before done a caesarean section as the surgeon, and no one else on my staff had more than a small fraction of my own limited knowledge. The patient certainly needed a caesarean section, but if there were to be an adverse outcome to either the mother or the baby, I was squarely in the crosshairs for the blame. I could not proceed without the support of my professional peers.

So I called a meeting of every physician in the hospital, explained the problem in great detail, and said that I believed a caesarean section was required. Unanimously and vociferously, "Oh no. She really has not had good labor! She will deliver the baby!" Under the circumstances, it was impossible for me to proceed. So I gave Anita intravenous fluids and penicillin, and I prayed! Another agonizing day and a worse night. But the next morning, there was a miracle! A ship had docked in Shanghai with the wives of every one of my medical officers except the anesthesiologist, and they had all left for Shanghai!

There was a medical officer attached to the MAG (military advisory group) whom I had never met. This was Col. Floyd Wergland, a general practitioner, graduate of Loma Linda Medical School, whose most recent assignment before China had been in the Office of the Surgeon General where he had been in charge of setting up the medical resident training programs for the armed forces. I called Col. Wergland, explained the circumstances, and said that I needed his help. "Get things ready. I will be right over!" I alerted the army nurse and the anesthesiologist (whom I shall call Smitty) to prepare the operating room. I spoke to Anita and her husband, explained the grave circumstances and the need for caesarean section.

I drew blood from Anita for typing and crossmatching, then I quickly assembled a roster of enlisted men to find a compatible donor and drew a pint of blood. I reviewed our stock of obstetrical pharmaceuticals and found that we had a single-dose ampoule of Pitocin (a drug which aids contraction of the postpartum uterus). One vial only! No further pharmaceutical help!

Then we were ready! Colonel Wergland and I exchanged introductions and scrubbed. Smitty started an intravenous line then put Anita to sleep. We scrubbed her abdomen, applied sterile drapes, and we were off! As operations went, cesarean section was pretty simple, and within a few minutes—through a lower segment transverse uterine incision—we had the baby out! A boy. We handed him off to Smitty, and the nurse began to slap his butt to stimulate him to breathe. Meantime, Col. Wergland and I extracted the placenta and horrendous hemorrhage ensued! The uterus, beat up by the prolonged, ineffective labor, was not contracting properly and thus was not constricting the many blood vessels which connect the placenta with the maternal circulation! And the baby was not breathing! I could see the prospect of both mother and baby succumbing!

"Smitty, is that baby breathing?"

"Not yet."

A few desperate minutes later, "Smitty, is that baby breathing?"

"Not yet."

Finally, I realized that there was no hope for the baby and stopped asking. The bleeding was furious. My one little transfusion was *pit-pit*ting away—miniscule against the tide of hemorrhage. We were madly packing the uterus to compress the blood vessels. Our one shot of Pitocin had been given, and we were reduced to despair.

"Smitty, *is that baby breathing?*"

"Oh yes, he has been breathing for quite a while!"

Instantly, Col. Wergland and I were friends for life—united by a common enemy! And miraculously, finally, that beat-up uterus began to contract and pinch off the hemorrhage!

On the next day, our fellow physicians and their wives flew in from Shanghai, and life moved on unperturbed. Both patients thrived. On the occasions of his birthdays, the parents sent me a photograph until he was fifteen—a handsome and happy young man. Not as happy as I, though! "Safe deliverance" indeed!

And shortly after, my own wife, Dorothea, arrived! She had received orders from the army to proceed from Boston to Seattle, thence to Shanghai and Nanking. My brother George had flown to Boston and driven our old Ford V8 Cabriolet to Post Falls. Later, Dorothea traveled by train to join the family until orders arrived for her to ship to China. Dad drove her on to Seattle and helped put the Ford on board. After an uneventful voyage, she and I were united and able to live with her parents in their home at 4 Shanghai Lu (Street)! Dorothea's father,

Dr. Stanley Smith—my buddy from Shanghai—was back at his regular job as vice principal of the Nanking Theological Seminary. Her mother, Dorothy, was busy with community service work (she formerly had taught music and home economics at Hillside, the Nanking American School). Actually, the old retinue of family servants was also on board, so this was a very happy interval!

On July 4, 1947, General George Catlett Marshall, United States ambassador, invited all American citizens in Nanking to the embassy. This was the only time I ever met General Marshall. He was a large man, but his head was large in proportion to his body. His face conveyed an instantaneous and unforgettable sense of great power and resolve but with an additional impression of interest in and concern for each individual whom he addressed. This was such an inspiring event that I was instantly convinced that, if I were ever to see the face of God, he would look like General George Catlett Marshall! I have that vivid impression to this day! He was, without doubt, the most remarkable human whom I have ever met.

There was some time for recreation. Dorothea and I made many friends among our fellow travelers—playing bridge, watching movies at the embassy, and various parties. At the embassy, we met a young American adventurer who was a renowned mountain climber and explorer: Bradford Washburn. He was in China on an expedition financed by Reynolds Corporation (the first ballpoint pen manufacturer) to climb the Himalaya's second highest peak, K-2. His plane had been damaged on landing in Shanghai, so he had an enforced layover waiting for repair parts from the States and made the best of his circumstances by coming to Nanking! He later became the director (and main motivator) of the Boston Museum of Science. We remained friends throughout the remainder of his life.

I found a good friend in Major Max Baughman, infantry, who shared my love of hunting. On a few occasions, we walked outside the Nanking City walls into the surrounding farmland. The fields were all small square or rectangular plots separated by dirt walls topped by narrow walkways. The main crop was rice, planted in spring when the plots were flooded by a few inches of water contained within these embankments. In the late summer and fall, they were dry and often had shocks of grain standing by waiting for threshing. We saw this process. A square, tightly-knit bamboo mat some fourteen to sixteen feet square was used as a platform to beat out the rice grain with a hinged (leather thong) bamboo flail. The recovered grain was stored in tightly woven bamboo baskets and transported across the shoulders on a bamboo pole. Housing in the countryside was in primitive small huts.

Although Max and I each carried a government-issued 12-gauge shotgun, I do not remember ever firing a shot, although there were many beautiful Chinese pheasants. We were apprehensive about the number of people in the fields—often obscured from sight by intervening grain shocks or other obstructions. On one occasion, we met a Chinese pheasant hunter who came toward us walking on the top of the embankment with a bamboo pole across his shoulders, carrying three pheasants dangling from either end, and a small net! He had, obviously, thrown his net over each one in succession and filled his quota! As we met, we exchanged greetings and hilarious laughter! The situation was preposterous: Two American hunters with the best modern firearms and no pheasants and one Chinese with a net and a bamboo pole and *six* pheasants!

This incident is reminiscent of our uniformly happy relationships with our Chinese hosts. The Chinese word for the United States means, literally, "beautiful country;" and for Americans, "people of the beautiful country." Americans were considered to be China's best friends. Japanese, Russians, and Germans were viewed extremely negatively. In all my dealings, I found our attitudes and outlook (Chinese and American) to be remarkably similar; in particular, we share a similar sense of humor and are quick to dissolve into laughter. I thought then, and I believe more firmly now that it was tragic for America to abandon China to Communism. Another huge swing in our fated century!

But in the midst of happy, busy days, my mind was turning to the next phase. When I left the MGH for military service, the chief of surgery, Dr. Edward D. Churchill, was still away, serving as the commanding medical officer for the Mediterranean Theater of Operations. The acting chief, Dr. Oliver Cope, had assured me that there would be a place for me to return to the MGH. However, Dr. Churchill, on his return, first appointed Dr. Cope to be director of the residency program, but later replaced him by the appointment of Dr. Frances D. Moore to this responsibility.

Although I knew both Dr. Cope and Dr. Moore well, I was not receiving signals anticipating my early and happy return! I shared these thoughts with my new friend, Dr. Floyd Wergland. "No problem. Join the regular army, and we will put you right back at the MGH on full military pay and allowances, and we will see that you are set on a career track of academic surgery!" And so I applied for a commission! As a part of the screening, I was required to take intelligence tests, psychological tests, physical examination, etc.—only available in Tokyo. Thus, a two-week leave and a trip to Japan!

The Enigma of Japan

I arrived in Tokyo just over a year after the Japanese surrender. It was a shock! Tokyo was a city of burned-out, rusting metal frames of burned buildings. Aside from American military vehicles, the streets were empty—save for a few trucks powered by charcoal burners sitting behind the cabs, requiring a pause, and firing up before attempting a hill. The streets were filled with pedestrians—each wearing a surgical-type face mask and carrying a small, round box containing a lunch, invariably fish and rice. Without exception, these people were courteous and *friendly*! How could this be? We are their enemies/conquerors! One could not avoid wondering, "Is this a ruse?" I think it was, in fact, genuine, but very puzzling.

General Douglas MacArthur was, at the time, *de-facto* regent of Japan. His office was an impressive stone building (with a façade reminiscent of the White House) directly across the street from the moat around the imperial grounds. On either end of the portico was stationed a military guard. Two of the tallest soldiers whom I ever saw marched back and forth at intervals in a manner reminiscent of the ceremony at the Tomb of the Unknown Soldier in Arlington National Cemetery. In comparison, the tiny Japanese visitors were effaced! Unmistakable language!

But the subways were running on time to the second! If you were in the doorway at take-off time, God help you!

I had some opportunity to travel in the countryside, to visit their delicate (paper?) houses, to visit warm springs bath, and to see an elaborate Buddhist temple. And I learned to use their latrines: not a simple accomplishment! The Japanese bodies are very flexible, being able to sit on their ankles for hours at a time apparently in comfort. Their latrines are not toilet seats (as our custom) but "mechanized" slit trenches with plumbing to flush. A major cultural adjustment!

In addition, I spent some time in Saint Luke's Hospital—at that time taken over by the American government. This was an impressive hospital even under wartime stress, and it is still worth contemplating the work of this institution which was founded in 1902 by Dr. Rudolph Bolling Teusler, a missionary doctor from the American Episcopal Church. In 1933, Dr. Teusler left this blessing: "This hospital is a living organism designed to demonstrate in convincing terms the transmuting power of Christian love when applied in relief of human suffering." Dr. Teusler's work goes on, enhanced, today. And I passed the examinations!

Completion of Military Tour

Back at work in Nanking a few weeks later, I received a letter from the United States Army Headquarters in Washington, DC, offering me a regular army commission. All that was required was for me to sign and return the acceptance. And *in the same mail*, I received a letter from Dr. Francis Moore assuring me that there would definitely be a place for me on the resident staff of the Massachusetts General Hospital when I returned from military duty! I returned my commission, unsigned, with thanks to Colonel Wergland.

In July 1947, I received orders to take Dorothea, proceed by air to Shanghai, board a Jardine-Matheson steamship, and sail to Hong Kong. There, I was to screen, for communicable diseases, a shipload of Chinese war brides prior to their transfer to the United States! During the war, many Chinese-American soldiers were shipped to western China, and many had found wives there. When the soldiers were shipped home, there was no place for women on the troopships, so the group had all been assembled for a more civilized voyage to their new country! I was given a two-week assignment and (upon my application) another five days to visit Canton (now Guangdong), fifty miles up the Pearl River from Hong Kong.

(The digression to Canton requires a little explanation. In July 1947, after my departure, Dr. Edward D. Churchill, chief of surgery, returned to MGH from his military assignment as commanding medical officer of the Mediterranean Theater. He brought with him several young men whom he wished to insert into the surgical residency program. Slots were scarce. He called in Dr. Chiu-an Wang and said to him, "Wang, you are going back to China after your residency, so you will not need your surgical boards. So you go along home." Faced with dismissal, Chiu-an could only take his bride and return to Canton. There, he was made chief of surgery at the Canton Hospital. When I was sent to Hong Kong, we seized the opportunity to visit them.)

In accordance with the orders, Dorothea and I flew to Shanghai and boarded the steamer. As we boarded, it was discovered that, as a captain, I was the highest-ranking military officer aboard and, therefore, entitled to cabin no.1 and a seat at the captain's table! So we rode in state to Hong Kong!

Hong Kong, even then, was a queenly city! Located on an island approximately a mile from the mainland, with a small territory on the adjacent shore for a suburb and an airport, it is squeezed into an area of thirty-one square miles (mostly a steep hill) for the city and four hundred square miles for the peninsula of Kowloon. Connecting the two areas is an efficient ferry system, and managing the mountain is a cog railway system. It has a large port, and occupying my attention was the USS *General Grant*—the passenger ship brought for my cargo! As it turned out, there were approximately six hundred women—none of whom were English speaking—who required screening for tuberculosis, parasites, venereal disease, and skin disease. This was a formidable job but was accomplished in ten days—just in time for the ship's sailing.

When time for boarding arrived, the captain insisted that I stand at the top of the gangplank with him. One by one, the women filed up the walk, carrying all varieties of bags and luggage. One woman had a large paper bag which caught the captain's attention. "What's in that?" he demanded. The poor woman understood not a word! Finally, by signs, he conveyed that he wanted to look into the bag. He took one look then demanded, "What is that?" Of course, *non comprende*! So he turned to me. "What is that?" Near the bottom of this large bag was a pile of brownish, gooey material which I certainly did not recognize. "Well, you can't bring it aboard!" Pandemonium! Not from the woman, but from the people around her-some of whom understood the captain's order. The crowd was absolutely resolute that this bag must go aboard! The captain was in something of a box. Finally, he said, "Well, you can't bring it aboard, but you can take it down to the post office and mail it, and I will hold the ship until you get back." I saw no more of the woman, but I presumed she made it back and boarded the ship.

(Seven years later, after I had finished my residency and served five years as a major in the US Army Reserves, I received a letter from the War Department: "Reply by endorsement to the secretary of war what you know about an incident on the Hong Kong Wharf when a Chinese woman was stopped for questioning as she boarded the USS *General Grant*!" Wow! The US government never forgets! I'm never going to cheat on my income taxes! It turned out that this was a packet of preserved bull's testes—meant for the comfort and encouragement of her husband!)

Dorothea and I traveled up the Pearl River to Canton aboard a small steamboat about a half-day trip through small towns and small agricultural plots planted mainly with rice. Chiu-an and Alice were living in a Quonset hut which had been placed on grounds adjacent to the hospital, and we had a joyful reunion in their home! Their first child, Irene (named for the great Hollywood set and costume designer [her work included the movie *Cleopatra*] who was Alice's friend) was seven months old and bedded in a crib standing on metal stilts about thirty inches tall (for protection from rats)! But all were well and in high spirits!

We had a very full (three-day) visit during which we met the "old aunt," had a celebratory feast, and met Chiu-an's youngest brother, Chiu-chen, who was an intern at the hospital. At Chiu-an's request, I did an operation on a six-year-old child to remove a six-centimeter bladder stone!

But there was also a pall of anxiety enveloping our thoughts and conversation. The Communist control of China was growing in territorial conquest and slowly coming down from the north.

Many of the Nationalist Party leaders were making plans to move to Taiwan if their territories were to be overrun. Both Chiu-an and Alice would be prime targets of the Communists because of their American connections. And things were tough economically. Before we left, I gave Chiu-an all the extra uniforms that I had packed, and he later told me that he wore them for years! As it happened in the end, as the Communists approached Canton, Chiu-an, Alice, and their children were able to board a plane and fly to Hong Kong in safety.

A Taste of Peking Duck

In early fall of 1947, Dorothea and I flew to Beijing for a few days of military leave. The people in that area seemed to speak the "purest" form of Mandarin—with a soft and musical accent that is so agreeable that one immediately knows that, if we ever get to heaven, we shall all be speaking the Beijing dialect! Those were days of poverty and hardship aggravated by the Japanese invasion in 1930 and the continuing Communist standoff from the mid-1920s. But still, we were greeted with warm hospitality and friendship. This is not the place for a travelogue of Beijing, but it is one of the most beautiful and significant cities in the world—in either prosperity or want.

A few anecdotes will convey some insight. Dorothea and I decided to have some Peking duck in Peking! So we were seated and our waiter seemed to get the message. Within a few minutes, we were surprised to find, driven around our table, a flock of ducks! We were required to select "our duck"! We chose appropriately and the ducks disappeared. Within a few minutes, each of us was served a bowl of duck soup! In leisurely style, the remaining feast came forth with suitable ceremony!

In imperial times, women—as infants—often had their feet bound tightly to restrain growth from early childhood to discourage their tendencies to "wandering" in more mature years. The feet on such women were unbelievably tiny—a few still living at the time of our visit.

Males who were to become court servants in the Forbidden City were castrated in their boyhood. A few of the imperial eunuchs were still living in a monastery in Beijing. They were a friendly, open, hospitable, and gentle group of men—all with fine-textured skin and hair!

Perhaps the most significant observation of our travels was the scholars' booths. These were small wooden huts about four by six by six feet in dimension, each having a door cut across the middle with a shelf to permit writing on the top of the lower half of the door. Upper and lower halves could be opened and closed independently of each other. In the imperial days, the ablest scholars from the entire empire were brought here for civil service examinations, and they lived in these huts during their testing. The brightest performers were appointed to provincial administration posts.

This system was in place from AD 605–1912. Was this the glue that held together the Chinese empire?

Peking Union Medical College

During our Peking visit, I spent a few days studying the so-called PUMC, the premier medical college of China—a joint project of the Chinese government, the Rockefeller Institute, United States government, and various United States medical schools. This institution had, in fact, been well-preserved through the Japanese War years. It came as a shock to me to discover that the PUMC Library was *much* larger than that of Harvard University when I attended there! Their medical records' system was superior to that of the Massachusetts General Hospital! However, functionally, at that time, the hospital was not doing much—probably reflecting the turmoil of the Communist revolution.

Summary Comment

Beijng is a city of remarkable beauty and grace, with a significant role in the formation and support of world civilization. Its spoken dialect is so soft and beautiful that it is my impression that, if we all get to heaven, we shall all be speaking the Mandarin dialect!

Other Assignments

In November 1947, General Marshall had returned to Washington and had been replaced as ambassador by Mr. J. Leighton Stuart. The wife of an American Presbyterian missionary on Hainan Island (in the South China Sea) fell and broke her right hip. A call for help reached the American Embassy in Nanking. It was presumed that she might need an operation for internal fixation of the fracture, and no surgeon capable of doing this was in Southern China. In fact, there were only two such surgeons in all of China: Dr. Williams Cochrane in Pao-ding-fu and me. Dr. Cochrane had been born in China, the son of missionary parents, and had become a Presbyterian missionary himself. He had two years of fellowship training in orthopedic surgery under Dr. Marius N. Smith-Peterson at the Massachusetts General Hospital. However, his city, ninety miles south of Beijing, was surrounded by Communist troops, and it was doubtful if a plane could get in and out without being shot down.

At seven o'clock one evening, I received a telephone call at home: "Rodkey, this is Gen. Lucas" (Major General John P. Lucas).

"Yes, sir!"

"At seven o'clock tomorrow, be at the airport, prepared to go to Hainan Island, and treat that woman's broken hip. Take whatever equipment and instruments you will need, and you may take an assistant."

"Yes, sir!" That was my total lifetime conversation with the general!

At 7:00 AM, precisely, we handed over our gear and boarded the C-47 which had been fitted with passenger seats. The plane was *filled*! We took the only two seats remaining! The rest seemed to be populated with majors and colonels! However, shortly after takeoff, we learned that we had two reporters aboard— one representing the Associated Press; the other, the United Press. These two gentlemen were seated conveniently in the row immediately in front of us! They were intent on getting the story of just what we planned to do for the afflicted lady! As it turned out, we had no significant conversations with the remaining passengers.

The reporters were persistent in pursuing the details of my proposed treatment. They were entirely unmoved by reminders that I did not know the woman, had no idea whatever of the extent of her injury, and many other similar minor details. They shifted tactics: Who was Dr. Smith-Petersen? What operations had he devised for treatment of fractured hips? Who was Dr. Williams Cochrane? What were all the possible ways of treating fractured hips in this era? It was a dreary five hours to Canton!

We stayed overnight at a hotel in Canton. Aside from the Broadway Mansions, this was the coldest night I ever spent in hostelry! The hotel had been designed to keep cool in hot weather—with the doors chopped off at knee height to allow the breezes to blow—and sheets, no blankets! A teeth-chattering, sleepless night!

On the following morning, we continued our flight to Hainan, arriving midday. As we circled, there was a stunning sight below—a huge airport! As it turned out, Hainan Island had been a major airport and submarine pen for the Japanese military forces, and we were the first American military forces to visit following the Japanese surrender! The presence of my fellow travelers was intuitively explained. They fanned out over the entire island the moment we landed!

My assistant and I were warmly received at the hospital; however, the x-ray equipment was down, undergoing repair and not expected to be functioning for another two days! Our patient was in a hospital bed, her right lower extremity in traction but reasonably comfortable. However, without an x-ray, it was impossible to plan her treatment! When our military contingent returned from their survey mission, the first question was "when will you be ready to leave?" There was great dismay at the news!

Within a few minutes, a contingency plan was proposed: "You don't need us here. We shall fly over to Hong Kong for a little R and R, and you can notify us when you are ready to leave." To this, I readily agreed, and my assistant asked, "Do you really need me, or may I go too?"

"Yes, you may go!"

Whereupon, the whole crew—including the two reporters—cranked up the plane and departed!

The emotional tension in the place calmed immediately, and I had some R and R myself! As it happened, the next day was November 17, 1947—my thirtieth birthday! My generous hosts gave me a dinner in celebration featuring a thousand-year-old egg (pidan)! This was the most peculiar-looking black egg! However, it had a pungent, pleasant taste. (Later, I learned that its life span was more like four months, treated with salt, tea, lime, and wood ash!)

The x-ray machine was repaired, and we took a single anteroposterior film of the patient's right hip. It was immediately apparent that she had a fracture of the hip, but with the femoral head impacted in valgus—that is, crushed into the neck with its ball pointed slightly upward. The one type of hip fracture that did not require an operation! All that would be required for healing was to keep her at rest with no weight bearing on the hip for six weeks! However, prolonged rest entailed significant risk of her developing a clot in the major veins in her leg that might fragment and float into her lungs—a potentially fatal complication. At that time (before the availability of anticoagulants), the accepted management of this problem was to tie off the major vein at the upper thigh. This I did, using local anesthesia. The lady recovered with no problems, but when she contacted her family in Chicago, she found that they knew all about the operation she had received for placing the Smith-Petersen nail in repair of her fracture! They had read it in the *Chicago Tribune* the day after the reporters went to Hong Kong for R and R!

There is an interesting postscript to this story. About ten years ago, the Chinese intercepted an American military plane and forced it to land on Hainan Island. There, they disassembled and studied its capabilities. Its crew was, eventually, released. This was, I suspect, the first American military presence on Hainan since our visit!

Taiwan Assignment

In January 1948, I was sent to Taiwan Island on an inspection assignment. I flew into the airport at Kaohsiung at the southern tip of the Island—second largest city and the major port of Taiwan. This was a city of heavy industry and major international trade but currently in serious decline in the aftermath of the Japanese surrender. My assignment took me by rail from Kaohsiung to Taipei—the entire length of the island!

Taiwan is a beautiful and fascinating country. Named by Dutch explorers "Formosa" (in Portuguese means beautiful island), the island was populated by aborigines. Some of these still survived in the eastern mountains and were reported, still, to be cannibals! It is about 250 miles long and 100 miles wide, with north-south orientation, and covered by five rugged mountain ranges over its eastern two-thirds. The climate is warm and humid, and there always seems to be a beautiful white cloud hovering high above. The western third of the island is beautiful plains, gently sloping down to the sea and covered with verdant green plots—apparently various types of fields.

Over a period of many centuries, western Formosa had been populated by Chinese. They were aggressive pirates, and in 1874, the Japanese retaliated by installing a military government. In 1895, as part of the settlement of the Sino-Japanese War over the control of Korea, Japan took sovereignty in Formosa. This status had continued until the Japanese surrender to Gen. Douglas MacArthur on September 2, 1945. As I traveled north, residual evidence of the Japanese occupation was everywhere—in the people, the language, the architecture, the signage, the roads, and the railroad. In fact, the whole island felt as if it were in a vacuum of uncertainty! And it was on the verge of a rush of Chinese Nationalist officials and their associates as refugees from the Communist conquest of mainland China.

Interestingly, in the midportion of the island, I saw an outpost laboratory of the Rockefeller Foundation that was dedicated to the study of malaria. Taiwan (Formosa) had especially vicious malaria problems with three varieties: *Plasmodium vivax, Plasmodium malariae,* and *Plasmodium falciparum.* The mosquito vectors that transmit these diseases have different preferred ranges in altitude, so the diseases tend to occur in altitude bands. However, some individuals were infected with all three varieties! Malaria was then—and remains—one of the most destructive disease scourges of the world. The disease is caused by blood-borne parasites of several closely related species of mosquitoes, which have complex life cycles spanning both their insect and human hosts' life phases. Despite strenuous efforts, no successful vaccine has been developed.

Taipei was a beautiful, modern city with an air of "pause before the storm." Everyone was in a state of anxiety and uncertainty, looking forward to the likely cataclysmic changes in government structure. The Japanese "hangover" was pervasive, not a comfortable time!

Chapter 13

TRANSITION

Times were changing for us too. The end of military service was in sight and orders had been cut. We were to fly to Hawaii, take a week's leave at Fort DeRussy in Honolulu, fly to San Francisco, and thence proceed to Camp Pendleton for separation from the army. We packed our gear, sold our Ford, and took sad leave of our friends and family!

Our flight from Nanking to Shanghai and thence to Honolulu was uneventful, and we were whisked into spring! Fort DeRussy is on Waikiki Beach, and that time apparently was chiefly devoted to rest and recuperation (R and R). This was a beautiful time of resting, exploring Honolulu and its beaches, wearing a few leis, savoring some luaus, and considering the prospects of home—thanks to the hospitality of the United States Army! We spent one day touring the Tripler General Hospital, then newly built, as the army's chief medical center in the Pacific region.

And then it was time for home! We flew from Honolulu to San Francisco in a two-engine C-47 converted to passenger seats—an eight-hour trip. At a certain point, we heard an unusual sound, looked out the window, and saw one of the engines feathering its propeller! Within a few minutes, the captain came on the intercom and announced that the engine was out of service, that we were twenty-minutes short of the point of "no return," and that we would be turning back to Honolulu! It was a long ride back!

Our Country, 'Tis of *Thee*

It is difficult to describe our emotions on return to San Francisco: the American flag never looked so beautiful, and the sound of English speech flowing around us was wonderful! We can never fully appreciate the wonder and beauty of our own home without an experience of separation!

At Camp Pendleton, I received my honorable discharge and accepted a commission as major, Medical Corps, United States Army Reserves. We were now civilians and free to move about the country!

Dorothea and I took a train to Spokane and spent a week with family in Post Falls. We had all been significantly changed by the experiences of the war years, though we remained a cohesive unit. Brother Lee had developed crippling poliomyelitis and was unable to perform military duty. John had served in the air force as a radio operator in the Troop Transport Command in North Africa and in the Mediterranean Theater. George had served as a medical corpsman on a naval destroyer. Lee was assistant professor of biochemistry at Harvard Medical School. John was teaching in the Spokane schools, and George was a medical student at Wayne State University. All were married. Elizabeth and Kathryn were in high school! Dad and Mother were middle-aged. Mother was still teaching in the Post Falls School System and giving piano lessons! Dad was still running the farm—much less efficiently without part of his work crew! Money was still tight!

During our years on the Colorado prairie, we heard the fame of the Mayo Brothers' Clinic. Occasionally, one of our neighbors was shipped to Rochester, Minnesota, for treatment of grave illness. On our railway return to Boston, Dorothea and I detoured for a three-day visit at Rochester for me to study the medical record, library, pathology department, operating suite, and administration of the clinic. This was an advanced, coordinated system with highly effective performance. In particular, I envied the location of the frozen section laboratory adjacent to the operating theater—an arrangement that I later urged (without success!) during a renovation of the operating suite at the Massachusetts General Hospital.

And back to Boston in May, I resumed my training, now as an assistant resident in surgery, since I felt that general surgery would give me a much broader scope in the treatment of disease than I should achieve by going back into orthopedic surgery. I was not among the inner circle of Dr. Churchill's residents, so I had the great good fortune of being assigned to the peripheral specialties as well as the general surgery rotations, including urology, neurosurgery, and six months at the Salem Hospital. These experiences equipped me to be effective in treating a wide range of disease, and they greatly expanded my skills in diagnosis.

Before shipping off to the army in 1945, Dorothea and I had sublet our apartment in Poplar Place—adjacent to the hospital—to an incoming Harvard medical student from Whitworth: Loren Gothberg and his wife, Eleanor. Upon our return, Loren was graduating! So the transition in housing was efficient and painless! Dorothea found work as a medical social service worker at the MGH and at Women's Hospital in Roslindale. Both of our salaries were at subsistence levels, but with rental at thirty-five dollars per month, we were able to scrape through!

The six months' assignment to the Salem Hospital in 1949 was significant. I was at an advanced stage of training. I had significant independent surgical experience in Nanking, and I came under the tutelage of Dr. Walter Phippen. Dr. Phippen was a graduate of Harvard Medical School in 1906 and had his surgical training at MGH. He returned to his hometown to become chief of surgery and to bring the Salem Hospital to a high level of performance, as ranked among peer institutions. He was a strong community leader and a devoted teacher. My experience blossomed under his guidance. I was at loose ends as to my future career. The Communists had cut off all aspirations of serving in China, and I had no other alternative plans. Dr. Phippen said, "When you finish, come back and work with me. I shall retire in a few years, and you will take over the practice." It was a generous and promising offer. I accepted.

In January 1950, I returned to the MGH for my final six months of residency training. There, I learned that Dr. W. Phillip Giddings, assistant to Dr. Arthur W. Allen, would be resigning to join the staff of the Veterans Hospital in Bennington, Vermont. Dr. Allen was the foremost surgeon in Boston, a man well-known to Dr. Phippen. When I learned of Phil Giddings' plan, I drove to Salem to discuss the issue with Dr. Phippen. I suggested that it would be wise for me to work with Dr. Allen for a year before settling in Salem. Dr. Phippen enthusiastically agreed. Dr. Allen accepted my application, and the plan was set.

Dr. Arthur W. Allen was a towering figure in the world of surgery. To avoid digression, I shall not attempt to describe him here but shall append an obituary note which I wrote in 1958. Suffice it to say, I moved smoothly into Phil Giddings' shoes and loved the work, the people, and the opportunities for learning.

After I had been with him for about six months, Dr. Allen called me into his office. "I think you had better stay on and become a permanent member of the firm." (His associates then were Dr. Claude E. Welch and Dr. Gordon A. Donaldson.)

"Well, Dr. Allen, you know that I have this arrangement with Dr. Walter Phippen."

"Yes, I know. You must go and tell him about this offer. He will say, 'That is a terrible thing for Jimmie Allen to do, and I am very upset with him! But on the other hand, it is too good an opportunity for you to pass up! You must go and do it.'"

I drove to Salem, sat down in the office with Dr. Phippen, and told him the story.

He was *very* disturbed! He said, "That is a terrible thing for Jimmie Allen to do, and I am very upset with him! But on the other hand, it is too good an opportunity for you to pass up! You must go and do it!"

And so it happened!

Chapter 14

FAMILY AFFAIRS

There followed eight years of prodigious work, wonderful friendships with fellow workers and patients, the beginnings of my own surgical practice, the selection of my secretary and office manager, Ms. Suzanne Graham Seckel, who would remain with me throughout life, and the growth of our family. With the help of Dr. Allen and Dr. Welch, Dorothea and I bought a house in Belmont (near neighbors to the Welch's on Belmont Hill). We had no children, and for several years, we tried to find a child for adoption. It was, at that time, a discouraging experience. Private, church, and state adoption agencies were very prescriptive in their demands upon prospective adoptive parents, and the process of applications, interviews, and rejections—sometimes scornful—was most discouraging. In the midst of this, in 1956, I had a call from Harvey Frazier in Spokane. "A woman in my practice is going to put her new baby [coming in June] up for adoption! Would you like to adopt the child?" Talk about manna from heaven! We did not stop to think of what human tragedy might have led to this decision.

Harvey made arrangements with a lawyer and a judge—not just any judge but Judge Ralph E. Foley, father of the Speaker of the United States House of Representatives Mr. Thomas S. Foley. The baby, a girl, somewhat premature and underweight, was born on July 26, 1956. She was required to stay in hospital for three weeks to gain weight and strength. Then Dorothea and I flew to Spokane to adopt her and bring her home!

We went to Spokane Deaconess Hospital to pick up our daughter, whom we named Cheryl Anne, then went to the courthouse with our lawyer. We entered Judge Foley's courtroom to find a hearing in progress. As we came through the door, Judge Foley looked up, saw us, and declared a recess in the proceedings. He invited us into his chambers behind the bench and began to question us. I was stunned by his questions—so totally different from any we had previously faced!

It took me a few minutes to understand his point! "What wonderful people are these to have come all this way to give this child a home. I want to know more about them!" I am still moved to tears by the memory of this man's goodness!

In December 1958, we were able to repeat the process when our son, John Mark, was born. In each case, we spent a few days celebrating with our family in Spokane and Post Falls. And until his death, each time we visited Spokane, we took the children to visit Judge Foley and experience again his wonderful love—a beloved friend and benefactor! During each of these visits (biannually in July, until the children graduated from high school), we made a point of visiting the families of my siblings and promoting friendship and mutual support among all our children.

When we left Nanking in May 1949, the Communist army under Mao Tse-Tung was moving into the northeastern provinces. By June, the Communist forces reached the north side of the Yangtze River—opposite Nanking. Foreigners, including Dr. Stanley and Dorothy Smith, were evacuated to Shanghai. There, Dr. Smith served as the acting minister of the Shanghai American Church. As the Communist army advanced on Shanghai, the Smiths were evacuated to Singapore. Dr. Smith began a term as principal of a mission school there, which continued until his retirement in 1958 when he returned New York City to work with his old seminary (Auburn) that had now merged with Union Seminary. He, an avid tennis player, had developed heart problems manifested by angina pectoris and limitation in physical activity. In July 1959, he died suddenly in their New York apartment of an apparent acute heart attack.

Dorothea was the only surviving child in that family, and we were blessed to have her widowed mother join us in Belmont. Mother Smith was a gracious and good soul and a treasured help in our lives and in the lives of our two young children, Cheryl and John. We had ten years of happy companionship until she too died of cardiac arrest. It was a most remarkable circumstance that I was present with both Stanley and Dorothy as they died!

When we moved from Boston to Belmont in 1952, one of my patients gave us a pedigreed young female border terrier named Toni. Toni was a cherished member of the family and provided us with several litters of pedigreed puppies whom we gave to close friends who also had young children. She was a great teacher and beloved friend for Cheryl and John, as well as for Dorothea, Grandmother Smith, and me! Beautiful, beautiful little friend—beautiful in form and beautiful in spirit!

Our home in Belmont was a happy base for school, church, and community service activities. Cheryl and John attended the Belmont Schools through the primary grades. Cheryl attended high school at Chapel Hill in Waltham. Thereafter, she took a course in health services at Community College in Spokane, Washington, and spent twenty years as an assistant in a large pediatric clinic in

Phoenix. Sadly, she developed Hodgkin's lymphoma and succumbed to pulmonary failure during her chemotherapy in February 2007. She lived a warm and generous life of constant and cheerful service to others—a gift of love.

John attended the Rivers School in Waltham, then Boston University, and the McCormick Theological Seminary in Chicago, becoming an ordained Presbyterian minister. He married Linda Krauss (of Belmont), and they have three children: Ashley Helen, an elementary school teacher of autistic students; Grant Edward, a psychology major, a computer guru, employee of Staples, electronic engineer, and a restaurateur; and Hope Catherine, art major and candidate art restoration specialist, business office manager, and continuing art student at University of Massachusetts.

Dorothea, from the time we returned from China, took a leadership role in developing an organized staff wives' group for Massachusetts General Hospital's medical and surgical residents. Gradually, this was merged into a support group for the families of the entire hospital staff—with particular attention to the families of trainees coming from other countries. This continued throughout her life. After her death, there was found in her files continuing correspondence with families living in forty-four countries external to the United States—all former graduate trainees at the MGH!

At the meeting of the Clinical Congress of the American College of Surgeons held in New Orleans in 2007, as I was having breakfast in a motel dining room, I looked across the room and noticed a couple who looked familiar. As I finished, I walked to their table and asked, "Aren't you Snorri Snorrasson from Reykjavic, Iceland?"

"Yes," he replied, "and you are Grant Rodkey from the Massachusetts General Hospital!" He had been a surgical fellow at the MGH during the 1950s, and his family had been encompassed in Dorothea's outreach. "Abroad," he said, "the Massachusetts General Hospital is famous for two things: scientific excellence and hospitality!" Dorothea was a moving spirit in the latter.

As an outgrowth of our work in the Plymouth Congregational Church in Belmont and in the early 1980s, Dorothea and I participated in organizing a chapter of Amigos de las Americas. Our son, John, was also involved and spent a summer in Ecuador with the Amigos program. Dorothea continued an active leadership role in this organization until her death in 2009—twenty-one days short of the sixty-seventh anniversary of our marriage.

Chapter 15

FLEDGLING DAYS

The launching of a young physician in the transition from trainee, resident, or fellow to licensed independent practitioner is a delicate, complex process, encompassing legal, spatial, educational, social, economic, and attitudinal adjustments of delicate issues that become more labyrinthine with each passing year of social and legal complexities—far from the old "affable and available" concept! In my own case, I was fortunate to make my debut under Dr. Allen's wing. He and his associates, Dr. Claude E. Welch and Dr. Gordon A. Donaldson had offices in a building at 266 Beacon Street in Boston—a building owned and occupied by Dr. Marius Smith-Peterson, my old chief on the orthopedic service. I was given a small independent office space and authorized to hire a secretary. Although I was permitted to have patients of my own, for many months, these were extremely sparse. My work was mainly to assist in the management of the operations and postoperative care of patients of my senior associates—a very busy assignment—and one with enormous opportunities for learning the fundamental skills of a premier surgical practice. This included, of course, the opportunity to develop friendships with our wonderful patients!

But I was also given some special assignments. A family of Dr. Allen's patients had a daughter who had grown into her teen years as a very troubled young lady. In fact, as I soon learned, she was a hippie long before the concept and the term were ever hatched! It was decided (among my seniors) that I, being nearer to the age of this tormented girl than anyone else in the group, should be assigned to the project of bringing her back to the land of the living! Far cry from surgery!

The young lady and I had a few meetings together—not altogether peaceable—with gradual achievement of more understanding between us. Then she dropped out, and I never saw her again. Many months later, in a stairwell at the Phillips House at MGH, I met face-to-face Father Whitney Hale, rector of the Church of the Advent and a commanding figure. He stopped me, saying, "Grant, I want to thank you on behalf of the [affected] family for what you did for their daughter."

"I don't know, Father Hale. I'm afraid I was not able to help much."

"Tut, tut!" He jammed his bony forefinger deep into my chest bone! "Don't belittle the work of the Lord!"

Am I going to forget that encounter?

I was blessed to have the help of some talented young women as my secretaries (a term now replaced by the more descriptive title "office manager"). They didn't stay long, though, being attracted by graduate school or matrimony! My second secretary, after two years with me, answered an AMA (American Medical Association) survey form as to my type of practice: "General practitioner with a tendency toward surgery"! In 1955, during the course of my visits to the radiology department to study films and confer with the radiologists concerning our patients, I was frequently assisted by a cheerful, efficient, and very attractive employee of that department working as a receptionist and filing clerk. She was a graduate of Simmons College, with a minor in science and a major in business administration. As my current secretary had given me notice that she would be moving on, I asked Ms. Suzanne Seckel if she would resign her position and become my secretary. After careful consideration, she consented, and we began a busy, productive, happy life working together—which is continuing. Sadly, her boss, Dr. Laurence Robbins, chief of radiology, became so angry that he never spoke to me again so long as he lived!

Chapter 16

MASSACHUSETTS INSTITUTE OF TECHNOLOGY (MIT)

AND

HALE HOSPITAL, HAVERHILL, MA

In the early days of working with Dr. Allen, my salary was not overpowering—$250.00 per month. In order to supplement this, I obtained an appointment to the surgical clinic of MIT to work one afternoon per week. The chief of surgery at that clinic was on the staff of the Sancta Maria and Mount Auburn Hospitals in Cambridge, and our patients from the clinic who needed operative or other hospital care were treated there. Thus, I became acquainted with those institutions and many of their professional staff members.

Our patients were the faculty and students of the institute, and our clinic staff included a nurse and a secretary. This was a pretty stimulating crowd! Our secretary, Ruth, a highly experienced medical secretary (whose husband was an antique dealer in Cambridge) was one of the most remarkable people whom I have ever met. She was able to remember the time, the date, and all details of any day (or fraction of a day) in her life, any paper she had read, or any letter or other composition which she had ever written! This included the time, date, and minute details of experiences with people around her. This effortless recall was eerie—if not downright intimidating! Who wants all his behavior indelibly remembered? She was certainly someone with whom one should keep a cordial relationship! However, she was a kind, helpful, and engaging person with tolerance of the foibles of her companions on life's voyage.

Ruth's remarkable talents have stimulated me to some investigation of memory. This faculty is called eidetic memory or photographic memory, and it includes the ability to recall perfectly images, sounds or objects, dates and times of conversations, full recall of content, and full recall of actions or events with time and date. There must be variations in this syndrome, as demonstrated by my aunt Margaret's perfect recall of all component elements of a musical composition. Another example was Franklin Roosevelt's postmaster general, Mr. James Farley, who could recall all personal and family data from anyone he ever met. This does not seem to be linked to other types of prodigy or genius so far as I have been able to discover by researching the limited information available. It is a very handy skill for a politician or his secretary (i.e., President Franklin D. Roosevelt who snatched up Postmaster Farley)!

Recently there has been scientific interest in studying this phenomenon. An *entre* to this review may be found in the article "Remembrance of All Things Past" (James L. McGaugh and Aurora LePort. February 2014 ScientificAmerican. com pages 41–45).

In later years, I had the sad responsibility of caring for Ruth's mother with terminal thyroid carcinoma.

Our clinic nurse was married to Prof. Samuel H. Caldwell, who had been trained by Dr. Vannevar Bush. (Dr. Bush's son, Richard Bush, had been one of my medical school classmates and one of my fellow house officers at the MGH). Professor Caldwell was an electrical engineer and one of the early developers of computers. They had a home in Watertown and were frequent generous hosts to our clinic staff.

Professor Caldwell adapted computers to the task of Chinese printing. Chinese characters are, in fact, pictograms made up of a series of strokes by a fine brush with black ink paint always proceeding from left to right and top to bottom in a prescribed, formal order.

The Chinese language is monosyllabic—that is, each word has an individual vowel-consonant combination, and these are never combined into compound words. There are insufficient vowel-consonant combinations to accommodate all human concepts, so the utility of these combinations is multiplied by using changes in tone (or intonation) for different meanings. These modifications vary from one region to another; thus, Mandarin (Beijing) has four tones; Cantonese has seven tones. However, the traditional written Chinese was composed by combinations of one or more of twenty-eight strokes—each one always entered in the same, precise numerical order—moving from left to right and from top to bottom. From ancient time, Chinese printing was done by inking lead plate configurations of these stroke combinations of words, then pressing the image onto the paper. Classical Chinese writing was universally uniform over the entire population and over the course of millennia.

Professor Caldwell computerized these strokes and designed a "printer" which would select the needed strokes in the appropriate sequence. He found, remarkably, that on average, four strokes indelibly select the character (or word); and he programmed the computer to print the character as soon as the critical determinant stroke was struck. Thus, this ancient language proved to have a mechanical advantage over English which, on average, requires five letters to determine the word!

An unforgettable clinic patient was Prof. Norbert Weiner. Professor Weiner was a frequent visitor, and we became pretty well acquainted. He had been a true child prodigy, born in 1894 in Columbia, Missouri, with special skills predominantly in mathematics. He graduated with a bachelor of arts from Tufts in 1909 and with a PhD from Harvard in 1912 (age seventeen). Thereafter, he became a professor in mathematics at MIT. He had not been involved in the Manhattan Project but had done basic work in the development of computers. He had served some period in China, and he loved to practice his Mandarin, which certainly was no better than mine—definitely not at genius level! His absent-mindedness was the source of many amusing stories. MIT had multiple buildings connected by underground walkways. On a certain occasion in this passageway, Professor Weiner and a student collided. The embarrassed student put out his arms and caught the professor, to whom he apologized profusely.

"That's all right. No harm done. But tell me, when we met, in which direction was I walking?"

The student, a bit nonplused, replied, "Why, professor, you were going in this direction," pointing the way.

"Good. Thank you. Then I have had my lunch."

On one day, when I was in the clinic, we received a frantic call from the wife of a student from their apartment in temporary surplus military housing on the western side of the campus. These were pre-911 days, so I ran to my car and drove quickly to their home. There, in the bathroom, I found the student covered with blood, vomiting blood, and passing bloody stools, pale as a ghost and barely conscious! No time for a history and physical examination! I put him into my car and drove straight to the emergency ward at MGH. There, he was given intravenous fluids and blood transfusions but continued rapid gastrointestinal bleeding. Within an hour, we had him in the operating room where we found massive bleeding from the gastroduodenal artery in the floor of a deep duodenal ulcer. We ligated the artery and did a partial gastrectomy and vagotomy. He survived to become a lifetime friend and a senior engineer for Raytheon, and he lived to age ninety-two!

In the days before I joined the Allen team, first Dr. Allen and later, Dr. Claude Welch served as consultant surgeon for the Hale Hospital in Haverhill, Massachusetts, located about twenty-five miles north of Boston. As a part of my junior duties, I was assigned to this position. I was fortunate to gain the confidence of a very well-trained (University of Pennsylvania) and experienced general practitioner named Dr. Nathaniel Miller, practicing in Haverhill.

It was not long before I was called to do a complex surgical case at the Hale Hospital. The patient was a seventy-year-old dentist—diabetic, overweight, having had a previous heart attack, with the misfortune of developing a cancer of the sigmoid colon. So I was summoned to do the case, which was scheduled at 10.00 AM on a certain date—my first solo operation outside the MGH since returning from China. I arrived to find that Dr. Miller would be my second assistant, and my first assistant was Dr. Paul Nettle, whom I had never previously met. Dr. Nettle was cordial but taciturn. The anesthesiologist put our patient to sleep, then we scrubbed up and began the operation. As soon as we entered the abdomen, it was evident that we had trouble. The sigmoid colon had flipped over to the right lower abdomen, and the cancer had invaded the small bowel and the right ureter! This meant resection and repair of the small bowel, resection and ligation of the right ureter (in those days the ureter was not reanastomosed in such circumstances), and resection and reanastomosis of the sigmoid colon—all to be done in a subpar patient to say the least! I could see the outcome. This patient was going to die—my first case in this community. My goose was cooked!

At a certain point, about one o'clock, as I was struggling deep in the pelvis, I was muttering repeatedly to myself, "It ain't easy, it ain't easy."

Dr. Nettle became annoyed. "Of course it's not easy. If it was going to be easy, do you think you would be out here doing it?" That put things into perspective! Our patient survived, and five years later, I was called to repair his inguinal hernia!

Operating upon seriously ill people in a facility with scarce resources brought the recognition that nature will support the patient if the surgeon adheres to physiological principles, does the minimum disruption of function, and is skillful in technique. "Tiptoe up to the table, gently breathe on the patient, then tiptoe away before his body senses that you were there!"

Dr. Paul Nettle was the most remarkable gentleman. He was a graduate of Tufts Medical School in the days when a student went directly from high school to medical school, and he had supported himself by sweeping up the sawdust from the floor of a famous Boston bar. He became—as did everyone else—a family doctor in the days when (as he described it) there was fierce competition among physicians. Most of the service was house call with horse and buggy. When a patient called the doctor, he or she called not just one but two or three doctors. The physician who arrived first took the case! Dr. Nettle kept his horse in the stall with a collar on and a harness suspended above so that when a call came, he could pull a lever, the harness would fall down on the horse, he would cinch up the hames, and be on his way at full trot!

He also described most physicians' offices as being rather long and narrow, door at one end, doctor's desk at the other, with a window behind him as he sat facing the patient. (No female doctors then!) After the patient's recitation of symptoms and the unhappy results achieved by consulting another doctor, one would say, "What medicine did he give you?" The patient would bring out her box of pills and hand them over. The doctor opened the box, studied the contents, screwed up a look of horror, stood up, turned around, opened the window, and tossed the pills out on the ground. "It's a wonder that poison didn't kill you!"

"Now that, Doctor, was competition!"

Dr. Nettle was chairman of the board of education in Haverhill (the Dr. Paul Nettle School functions to this day). He drove an old pickup Ford truck with his dog riding in the back. His dog always made hospital rounds with his master! But considering the paucity of his training, Dr. Nettle did a remarkable job of staying out of trouble! He was especially skillful at setting fractures and managing minor trauma. And he had a store of unusual knowledge. One day as we were operating, he remarked, "This would be a good case for a horsehair drain!"

"Horsehair drain! Do you really mean a drain made of the hair of a horse's tail?"

"Yes."

"How did you sterilize it?"

"Boiled it."

How could this possibly have been of any use? I tried to find some reference to this in the surgical literature—no trace. Finally, I had an inspiration: it worked by capillary attraction! Normally, we think of capillary attraction operating inside the lumen of fine tubes, and this is the force that lifts the sap of trees and gives us maple syrup. But capillary attraction also works in the interstices of the exterior walls of fine cylindrical strands that are closely packed together. Bingo! The horsehair drain! I could mimic that by taking zero chromic catgut, folding it to have four finely packed cylindrical (*sterile*) strands!

All wounds heal by following the same pattern: cut, hemorrhage, clot, inflammation with mobilization of fibroblasts, white cells, sprouting capillaries, tissue enzymes, epithelial coverage, etc. But there is a critical early time of accumulation of fluid in the wound which is out of the circulatory system and is more vulnerable to bacterial infection. Despite all precautions we can take, bacteria still are everywhere in our environment—external, internal, and in wounds. This stagnant wound fluid (serum) becomes an ideal bacterial culture medium being incubated at 37 degrees centigrade—just what bacteria love most! If I could conduct that fluid away in the early days of wound healing, perhaps it would be safer.

I began to use horsehair drains in all major wounds—so-called clean, clean-contaminated, and contaminated. I would leave the drain in place until serous fluid stopped flowing up out of the wound into the dressing (usually, forty-eight to seventy-two hours) then cut off one end and draw the strings out the other. This practice, added to my already stringent technique, essentially ended surgical wound infections for me for the rest of my career. Thank you, Dr. Nettle, for this benefit to our patients!

Incidentally, my great friend Dr. Nathanial Miller (general practitioner in Haverhill) referred to me many extraordinarily complicated surgical cases over the next several years—all of whom did well with good functional recovery. And he also taught me an important life lesson: "The tears of gratitude dry quickly!"

Chapter 17

PROFESSIONAL GROWTH

The decade of the 1950s brought the realization that talc powder on surgical gloves was an inflammatory irritant in the peritoneal cavity, which contributed to the formation of dangerous peritoneal adhesions. In response to this information, surgeons gradually developed and adopted surgical gloves that had no talc. During this period of study and debate, I began snipping out fibrous adhesions encountered in the abdomen and sending them for microscopic examination by the pathologists at the Massachusetts General Hospital. Talc particles were present in many of these specimens, but—to our astonishment—textile fibers (specifically, cotton staples, which are the lint shed from all cotton fabrics) were much more frequently the apparent cause of adhesions.

In the 1960s, I was appointed as the MGH representative to the Ethicon General Surgery Advisory Panel. I explained these findings to Ethicon and asked them to develop surgical sponges that would not shed such staples. Within two years, they developed surgical sponges that were absorbent textiles made from fabric created by injecting a fine stream of cellulose into a chemical bath to precipitate a thread. They called this product Sof-Wick. I immediately began to use it exclusively in all surgical procedures, and I educated all my fellow surgical staff members at the MGH regarding the issues. Of them, only one—Dr. Ronald Malt—took up my suggestion. Immediately, it became apparent that operative procedures done with this technique developed fewer adhesions! There were many occasions to confirm this observation in patients who legitimately required two-stage operations as part of required surgical care. Also, I learned that individuals who had extensive adhesions from prior operation and who required lysis of adhesions as part of the current procedure all developed more fever, ileus, and abdominal pain postoperatively than those patients with only an initial operation in which Sof-Wick was used. This suggested that releasing the encapsulated cotton staples from peritoneal scars triggered a new inflammatory cycle.

Dr. Malt used this product until he stopped doing active surgery in 1988, and I also until I retired from MGH and moved to the VA Boston Hospital in 1992. I was unable to persuade the Veterans Administration to purchase Sof-Wick. Thereafter, MGH ceased purchasing the product, and Ethicon sold it to another manufacturer.

Postoperative abdominal adhesions remain a very serious, incompletely addressed problem in surgery which takes a huge toll in terms of patient mortality, morbidity, disability, and added health-care costs. Significantly, laparoscopic surgery (performed without cotton sponge contamination of the peritoneal cavity) has been shown to produce markedly fewer adhesions than comparable open operative procedures in which cotton sponges have been used. It is my profound conviction that this issue needs study on a large scale by means of a large, randomized clinical trial in order to overcome entrenched physician prejudice/dogma. This cannot be done by an individual surgeon because of professional bias and medicolegal risk. I know that the results will bring great benefits to our patients and reduce costs to the health-care system. The VA Health Care System is ideally positioned to do this research. I hope that these comments will motivate this critical technical advance in surgery by my colleagues at the Department of Veterans Affairs. This is a deplorable example of the "Semmelweis syndrome"—patient hazard secondary to physician obtusity.

Chapter 18

Transition Phase

The early 1950s marked a change in the relationship of hospitals with its professional staff members. Hospitals began to construct conjoined office buildings and to solicit physicians as paid or contracted employees. Our own practice moved to accommodate this trend.

Dr. Allen had taken on other associates: in 1952, Dr. Glenn E. Behringer; in 1954, Dr. Stephen E. Hedberg; in 1956, Dr. George P. Richardson; and in 1958, Dr. John F. Burke. Thus, each of us had the opportunity for mentored, graded, continuing training in the art and science of surgery.

In 1955, Drs. Welch, Donaldson, and Behringer moved from our offices at 266 Beacon Street to the newly constructed Warren Building on the MGH campus, although we continued to function as a closely knit team. Our daily routine was early morning operation, rounds as a group on each of Dr. Allen's patients, intensive rounds on each of one's own patients; and office visits in the afternoons. Shoehorned in were surgical grand rounds, rounds on the ward service (patients on our months of ward teaching assignments), and consultations and visits at outlying hospitals (Hale, Mount Auburn, Sancta Maria, Martha's Vineyard). Our secretaries played a vital role in scheduling, monitoring, communication, and patient relationships.

In addition, we physicians all had participating roles in the Massachusetts Medical Society and in various surgical organizations. Dr. Allen had been president of the Massachusetts Medical Society in 1948, of the American College of Surgeons in 1950 (and chairman of the ACS Board of Regents in 1951), and president of the Pan-Pacific Surgical Association in 1956. He had also become president of the Boston Medical Library in 1954. This venerable and precious scholarly resource represented the collective libraries of major Boston physicians since 1814—including that of an earlier library president, Dr. Oliver Wendell

Holmes. The library, located at 12 Fenway, was in dire financial straits, and Dr. Allen had been elected president to lead its restoration! I remember vividly being called into his office.

"Grant, you are going to give $600 to the Boston Medical Library!"

"I am?"

"You are!"

In those days, this was not peanuts! It certainly got my attention! And the library has commanded my interest through all the subsequent years. In 1964, it joined the library of Harvard Medical School to create the Countway Library of Medicine—the largest private medical library in the United States. Of all the Countway Library's holdings, the Boston Medical Library contributed about three quarters of the volumes and the entirety of its rare books collection. For a number of years, I also served as its president, and I continue an active advisory role.

The Massachusetts Medical Society continues to participate in the management of the Countway Library by providing four members of the BML trustees, as well as sustaining funds as permitted by earnings on its financial endowment. This direct connection between the Massachusetts Medical Society, Harvard Medical School, and the Countway Library provides special opportunities for the physicians of Massachusetts to have bedside or office access to current medical information. In more recent years, the National Library of Medicine in Bethesda, Maryland, has—through its PubMed program—greatly broadened public access to medical information and research.

Incidentally, Dr. Allen also went to New York City to discuss these issues with the manufacturer of a popular patent medicine for indigestion, "Bellands." He secured the promise of a bequest from Dr. Dennistoun Bell which remains to this day the largest component of the endowment of the Boston Medical Library!

In early 1958, Dr. Allen became gravely ill. He had required an abdominal operation in the 1940s by his colleague, Dr. Leland McKitrick. Some sort of a lymphoma tumor was excised and followed by x-ray therapy. He never referred to this experience except for the dry remark. "We should never be too proud to eat at our own trough!" But now he had developed a recurrence or a new primary lymphoid malignant tumor that was accompanied by hemorrhagic manifestations, including severe, frequent epistaxis, and bladder hemorrhage. Despite the pain and frequent hemorrhages (which had to be frightening), he remained positive and cheerful in his attitude—more concerned for the welfare of his associates than himself. "You must remember me as one who did his best work after he developed a fatal disease!" His death was mourned internationally—particularly in England and Scotland—and by thousands of his patients and friends. As one wrote, "His name was Arthur, but he was called Jimmie—Jimmie for short, and Jimmie for love." It is not possible adequately to categorize the greatness of this giant of medicine and of humanity. This is more adequately expressed in an obituary note from the Harvard Medical School Alumni Bulletin, which shall be attached here as an appendix (Allen Obituary).

After Dr. Allen's death, we remained at 266 Beacon Street until our future should become clearer. In 1955, Dr. Chiu-an Wang had come to Boston for a one-day stop on his return journey from Toronto to Hong Kong. He had come to Canada to take the "Empire Board"—an examination for medical proficiency that, if one passed, would qualify the candidate to practice medicine at any site within the British Empire. This immediately telegraphed to me the insight that Chiu-an was worried for the safety of his family if the Chinese Communists were to take over Hong Kong. In planning our future moves, Chiu-an's needs had to be incorporated.

As it happened, the Boston residential section on the northerly side of MGH—the so-called West End—had been taken by eminent domain by the city. All the buildings had been torn down and the entire section was being converted to high-rise apartment buildings. The building nearest the hospital (just across Fruit Street) was partly dedicated to medical office space. Before the building was completed, we contracted for a space on the second floor facing the hospital, which should provide adequate space for three physicians and their staff. We actually were able to direct the design for an attractive and highly efficient arrangement for physician offices, examining rooms, corridor laboratory and sterilizing space, comfortable secretarial spaces, a common waiting room, and a staff utility room. As it developed, the space was not only adequate but beautiful as well. We moved in 1960, and it was our home for eighteen years!

Initially, we had Dr. Stephen Hedberg and his secretary, followed by Dr. Robert Shaw and his secretary, but the space had really been designed for Dr. Chiu-an Wang. In 1960, Dr. Welch went to Australia on a lecture tour. He was aware of the history of Chiu-an's dismissal by Dr. Churchill and was inclined to agree with my suggestion that we should bring him on as a surgical fellow with our team for one year to help him renew his skills. En route to Australia, he stopped in Hong Kong and interviewed Chiu-an. This went well, so we arranged for him to come in July 1961 for a year's appointment. Chiu-an rented his Hong Kong house, brought his family, and rented a home in Lexington for the year.

The experience was not easy for Chiu-an. He had been speaking and writing Chinese nearly exclusively since 1945, and both his spoken and written English had become pretty rusty. The pace of medical development in Boston far outstripped his learning opportunities in Canton and Hong Kong. It was a tough year playing catch up!

Near the end of his fellowship year, I reminded Dr. Welch that the MGH had not allowed Chiu-an to complete surgical training requirements for the American Board of Surgery. I suggested that we should try to give him an appointment now, which would provide an opportunity to make up the deficiency. Dr. Welch consulted Dr. Churchill (surgical chief), and they came up with a firm negative response to this proposal.

Then I turned to Dr. Oliver Cope. "Dr. Cope, you know this story. The MGH owes to Dr. Wang an opportunity to complete his surgical training. Dr. Churchill will retire on July 1, and he will be succeeded by Dr. Paul Russell. You have influence with Dr. Russell. I propose that you ask Dr. Russell to appoint Chiu-an to the staff. You take him on to work in your laboratory and your surgical practice, and I will provide him with an office and secretary and pay him a salary."

"Good idea!"

And so it was done! On the last Sunday in June, we had a large farewell party for the Wang family in Dr. Welch's backyard in Belmont—feasted and shot off firecrackers! Then Chiu-an sent his family to Maine. I took him to the airport to fly to Hong Kong to sell his house, but only he, his wife Alice, and I knew that he would return next week! Dr. Churchill retired, Dr. Russell appointed Chiu-an to the staff, and we were off! When Chiu-an returned from Hong Kong, he brought me a Patek Philippe wristwatch that I am wearing still fifty-five years later!

But the struggle was far from over! Chiu-an, rusty in the English language, now had to pass his American Board of Surgery examinations. He studied hard for a year, took the examination, and failed. The same story the next year. Finally, on his third try, he passed!

When it came time for the second (the oral) part of his third examination, he was required to go to New York City. There, his examiner turned out to be Dr. John Mulholland, a prominent New York surgeon, expert in both gastrointestinal and endocrine surgery. Dr. Mulholland spent the entire hour in querying Chiu-an on his opinions regarding various aspects of endocrine surgery—particularly on parathyroid surgery! By this time, Chiu-an had done extensive postmortem and clinical studies on parathyroid disorders and was well on his way to becoming one of the world's foremost experts in the field of endocrine surgery.

We continued our close relationship and our affiliation with Dr. Cope. Chiu-An became one of the most renowned endocrine surgeons in the world—particularly in the field of parathyroid surgery. Chiu-an and Alice returned to China for a visit. During the year, he served as president of the Society of Endocrine Surgery. His presidential address was devoted to the changes which he had observed in China during his lifetime—a most interesting and insightful treatise (Surgery, Vol. 100, No. 6, pp. 943–947, December 1986). He retired as clinical professor, emeritus in Harvard University.

Chiu-an and Alice had a summer home on Martha's Vineyard. Unfortunately, there, he became infected with babesiosis, which slowly brought him down to his death in October 1996. He lies peacefully on a small hill overlooking a pond in Mount Auburn Cemetery in Cambridge.

Dr. Chiu-chen Wang, the younger brother whom I met in Canton, was sent by his elder brother to an internship in medicine in Syracuse, New York, in 1948–49. Thereafter, Dr. Chiu-chen Wang trained in radiology at the Massachusetts General Hospital in Boston. He joined the staff and faculty at MGH and Harvard. He became the world's foremost radiotherapist in the management of head and neck cancer. He died on December 21, 2005, at the age of eighty-three, leaving a daughter who is an anesthesiologist in Florida.

But the story is far from over! The Chinese catastrophe (five daughters, no sons for Chiu-an and Alice) has become a brilliant constellation in American society: teachers, actors, finance officers, artists, pilots, administrators, physicians, researchers—with the list still growing. The glory of America!

Chapter 19

Busy Clinical Years

Our practice at Zero Emerson Place grew quickly in large part because of the support of our patients and referring physicians. Our staff (Ms. Suzanne Seckel, Ms. Sheila Smith, Ms. Alice Luongo; and, in succession, Ms. Claire Canapary, RN; Ms. Georgianna Kachadoorian White, RN; and Ms. Anne Maker Gavin, RN) were extremely efficient and devoted to patients' welfare, and patients and their referring physicians responded with enthusiasm! We had a high volume of patients with an extremely complex spectrum of surgical problems and gratifying outcomes of care. Most minor surgical procedures were done in our office with local anesthesia. (Our nurse was told by the pathologists that we submitted more specimens to the MGH than any other surgeon and more than any institution except the Emerson Hospital in Concord!) And each patient was a particular individual—a friend, a fellow traveler along the road, and a teacher. Each of these friends had an important, particular story of life. A few of these stories will be included as appendices, but each was of equal importance.

During my year as chief resident in orthopedic surgery, Dr. Marius N. Smith-Petersen had given me for Christmas a half-inch bone chisel. I thought, *What a wonderful idea. When it is possible, I shall try to do that.* Beginning in the early 1950s, I began the custom of giving to each resident who had served a rotation term with me during the preceding year a Christmas present of a pair of Mayo-Stille-type six-inch dissecting scissors with a special design for fine dissection—the style that was favored by Dr. Arthur W. Allen as the very best dissecting scissors in the world. One of its main virtues is its short (four by two inches) handle/blade ratio, which permits optimum accuracy and speed in tissue dissection.

In 1990, the Stille Company called me from Sweden to say, "Dr. Rodkey, you have bought more of these scissors than any other individual in the world!

We want to send you a token of our appreciation!" In due course, one of their representatives came to our office with a pretty little case enclosing a gold-plated instrument and with a plaque engraving: "Presented to DR. GRANT RODKEY for His Recognition of Fine Instrument Quality." According to our calculations, we had bought approximately five hundred pairs! I never attend a surgical meeting anywhere that someone does not come up to me with some variant on the comment: "Grant, I still have those scissors you gave me, and I am using them every day!" Happy reminders of the scores of wonderful, talented, and dedicated young people with whom we have served!

Chapter 20

THE BROADER SCOPE OF
MEDICAL PRACTICE

One cannot serve as an active physician without quickly recognizing the expectations and restrictions upon our activities imposed by the various laws and social attitudes relating to medicine. These are constantly changing in response to social pressures, economic changes, and scientific discoveries. By reason of their special training and insight, physicians' contribution to such social discourse is an important element of leadership.

In 1948, I took the examination to be accredited by the Massachusetts State Board of Registration in Medicine. I was assigned license number 22028 (still my identification) and joined the Massachusetts Medical Society. In 1951, I took the certification examination for the American Board of Surgery (no. 3442) and was recertified by voluntary examination in 1996. I joined the American College of Surgeons as a fellow in 1952.

Very quickly, it became evident to a young physician that there were problems with the accessibility, quality, and cost of medical care for American citizens. Although my entire career was conducted with conscious concern for these issues (Every patient is entitled to the benefit of all my knowledge, skill, and office resources for treatment on the first visit and for a modest fee.), it soon became apparent that an individual acting alone cannot accomplish much in the way of social reform. Thus, in 1955, I came to a meeting of the Suffolk District Medical Society (held then at the headquarters office of the Massachusetts Medical Society at 8 Fenway, Boston) and announced: "I am fed up with belonging to a society which is always reacting, never advancing positive initiatives!" Whereupon my friends said, "Please, sit down. Be our guest!" Thus began a lifelong tussle!

When the Massachusetts Medical Society was chartered by the legislature in 1781, one of its precepts was "to do all things as necessary for the health and welfare of the citizens of the Commonwealth." Acting in accordance with this mandate, in the decade of the 1930s after prolonged debate, the society founded Blue Shield of Massachusetts. There were two programs: Plan A for individuals whose income was less than $5,000 per year and Plan B for those with higher incomes. Physicians were to make no additional charge for treatment over the Blue Shield allowance for individuals insured under Plan A; those with Plan B might be billed an additional amount above the plan allowance. To protect the fledgling corporation, only physicians who agreed to these terms by signing the "Participating Physician's Agreement" were allowed to treat Blue Shield–insured patients. Very rapidly, Blue Shield expanded to cover the majority of Massachusetts's citizens. Soon, dissatisfaction with Blue Shield allowances for physician services began to be evident, and high interest in the board of directors of Blue Shield followed. New members of this board were elected (by the board itself) from a list of candidates proposed by the committee to nominate Blue Shield directors of the medical society—of which I quickly was elected to be chairman. The board of directors was a distinguished and public-spirited body. A majority were laymen representing mainly business (employer's) interests but also including representatives of labor and the general public. In 1970, I myself was elected to the board, and was chairman from 1972–1977 when I resigned to become a candidate for president of the Massachusetts Medical Society. Despite strenuous efforts, it was not possible to resolve the tensions among employers, unions, the state government, and physicians over the issues of cost, quality, access, and monopoly control of health insurance by Blue Shield of Massachusetts. A group of physicians brought suit against Blue Shield because of its monopoly control of health insurance within the commonwealth.

I was, in fact, elected and served as president of the Massachusetts Medical Society during 1979–1980. During my term, considerable energy and money were expended in an attempt to resolve these political, economic, and legal issues but unsuccessfully. The issues came to a climax in 1981–1982 during the presidency of Dr. Stanley Wyman. At Dr. Wyman's request, I became the chief plaintiff for the society in its lawsuit against Blue Shield for its monopolistic control and restraint of trade. The trial was in federal court before Judge Andrew Caffrey who ruled in favor of the society. In his comments, Judge Caffrey made the significant observation that physicians' services are not fungible; that is, no two physicians' services are precisely similar. Unfortunately, on appeal, the federal appeals court (Justice Stephen Breyer) overturned Judge Caffrey's decision, ruling that Blue Shield was acting "as a parent to protect its minor children," and that its monopolistic control is legal. The Supreme Court declined to hear an appeal of this decision.

Chapter 21

THE AMERICAN MEDICAL ASSOCIATION

In 1973, I was elected alternate delegate from Massachusetts to the American Medical Association. This organization, founded in 1847, has representation from every state and territory of the United States; its membership is highly capable, knowledgeable about the social and economic—as well as public health—issues in each geographic area, and highly motivated to public service. In my experience, the house of delegates of the AMA is the most effective parliamentary body in American society. If the Congress of the United States had similar qualifications and motivation, this would be not only a different country but a different world!

I became a full delegate in 1978, and in 1982 was elected to the AMA Council on Medical Service, dealing with issues related to the cost, quality, and availability of medical services to the American public. From 1989–1991, I served as chairman. This remarkable body was staffed by a group of highly competent and brilliant AMA employees. The council members were all experienced clinicians, and its agenda was the protection of the quality, availability, and reasonable cost of health-care services for the American public. Recommendations of the council were forwarded to the house of delegates and were often translated into AMA policy and action. We were a band of brothers (and sisters!) immersed in the public health service of the American people.

In 1991, the board of trustees of the AMA appointed me as the founding chairman of the Relative Value Update Committee. The need to establish such a study group arose from initiatives from the federal government to standardize and rationalize payments for physician services in the Medicare and Medicaid programs. During the preceding decade, studies related to this imperative had been carried out at the Harvard University School of Public Health by Drs. William C. Hsiao and Jack Langenbrenner in an attempt to define valid input factors to the calculation of such a payment system. The Hsiao studies became a foundation for the functions of the Relative Value Update Committee (RUC).

The board (Trustees of the AMA) had been extremely thorough in their planning for the RUC. The general organizational outline included a central committee of twenty-five members representing the major medical associations/ societies, including the American Osteopathic Association, the American Academy of Pediatrics, and the American Academy of Child and Adolescent Psychiatry; an observer/member from the federal government (Health Care Financing Administration [HCFA] and, later, Centers for Medicare and Medicaid Services [CMS]); an advisory committee of seventy additional medical societies' representatives; the Current Procedural Terminology Committee of the AMA; and the RUC Health Care Professionals Advisory Committee with delegates representing physician assistants, nurses, occupational and physical therapists, optometrists, podiatrists, psychologists, social workers, audiologists, speech pathologists, and chiropractors.

The staff members selected by the board to serve the RUC were among the most highly talented AMA employees, led by Ms. Sandra Sherman and her associate, Ms. Sherry Smith—brilliant individuals, knowledgeable of the subject matter, and possessed of outstanding leadership skills.

With this meticulous preparation, the first meeting of the RUC was in late fall 1991 in the huge ballroom of a downtown Chicago hotel on a cold, dark day with at least one (mostly two or more) vacant seat(s) separating each delegate from another. No one knew anyone else, and no one wanted to know anyone else! I consider it one of the marvels of human society that the members of this group have grown to respect, admire, and even love each other!

At the initial session, members of the staff handed out colorful caps carrying the logo "RUC" for use by the Central Committee members during meetings. The significance of this gesture was to emphasize the judicial responsibilities that each member assumed. Explicitly, RUC members are charged to view objectively the effect of each issue upon the public welfare and the medical profession as a whole. Specifically, in considering RUC matters, no member is to act as an advocate for his own specialty but in consideration of the consequences of our actions on the public and on the entire profession of medicine. This set the tone for prime support of the public interest in all our deliberations—as well as promoted the solidarity of the group

One of the first major tasks of the RUC was to set up a "Relative Value Scale" for all the services provided throughout the specialties of medicine. No such valuation existed because of the restriction of antitrust laws. However, because the federal government was an integral component of the RUC, it was legal for us to accomplish this. Studies of every service or procedure were initiated, involving all parties who performed these particular services by using practice surveys, specialty surveys, and group discussions. Evaluation included consideration of time required, technical complexity, resource cost, and iatrogenic risk to the patient. There are many, many procedures, and the task was extremely difficult but was accomplished during the first year of the RUC.

Years later, at an RUC reunion, Dr. Ray Stower (past president of the American Osteopathic Association) asked, "Grant, do you remember the name of the first resolution committee you appointed me to chair?"

"No, what was it?"

"The Committee to Value Sperm Washing. I put it in my CV!"

And so for some eight thousand CPT procedure codes! The task is not likely to be repeated soon! However, each five years, the RUC does a review of the entire inventory.

Even a gross attempt to value this public service enterprise of organized medicine over the course of the past twenty-five years is staggering. A conservative estimate of the cumulative amount since 1991 is well over $175 million. And this neglects entirely the costs to the AMA of the CPT (Common Procedural Terminology) System which undergirds the entire Medicare/ Medicaid classification of physician services.

The RUC is continuing its highly effective role in health systems dynamics in the United States under the skilled direction of Ms. Sherry Smith, director, Division of Physician Payment Systems, American Medical Association.

Chapter 22

MORE COMPLEXITIES OF THE
UNITED STATES HEALTH-CARE SYSTEM

My presidency of the Massachusetts Medical Society during 1979–80 overlapped the term of President Richard Nixon and was a highly turbulent time in the debates on health policy in the United States. Paul M. Ellwood Jr. a student of Kaiser Permanente and other West Coast health-care foundations, convinced (then) secretary of Health and Human Services, Elliot Richardson, that such more tightly organized systems of care held the promise of both decreased cost and standardization of the quality of medical care. This led to the passage of the Health Maintenance Act of 1973, which required employers of twenty-five or more people to offer a federally certified HMO option for health insurance. Among other provisions, physician organizations became eligible to submit proposals to be recognized as HMOs if at least 250 physicians participated. For Massachusetts citizens and physicians, this was an opening to diminish the monopolistic control of Blue Cross Blue Shield.

We discovered that two physicians—Dr. Louis Alfano and Dr. Robert J. Brennan—had already incorporated such an organization, which they had named the Bay State Health Care Foundation. Physicians in Middlesex North and Middlesex Central District Medical Societies in Massachusetts were eligible to join. This was the perfect name! Why reinvent the wheel?

I contacted Drs. Alfano and Brennan and suggested enlarging the scope of the foundation to include physicians from all the district medical societies in the Boston metropolitan area. They readily agreed. However, launching this program was vehemently challenged by hospitals. In particular, a proposal was advanced to create the "thirteen teaching hospitals' (Boston) HMO." The competition raged hotly until, at a Christmas party, an executive of the (then) Peter Bent Brigham

Hospital asserted: "The physicians of Brigham Hospital will join the teaching hospitals' HMO on pain of losing their hospital privileges!" Within a week, the teaching hospitals' HMO was dead! Word of mouth is like a raging forest fire!

The Bay State Health Care Foundation was launched in 1970. Because of my close involvement with its founding, I never accepted a managerial office in the organization, although I kept very close surveillance over its affairs. There was an important opportunity and great public need for the work of this foundation; however, the administrative insight and skills that physicians (as trustees) brought were too naive to cope with ineffective and dishonest management shysters who brought the organization to bankruptcy and takeover by Blue Cross Blue Shield in 1993.

This was a significant eruption in the volcanic churning of health-care policy that actually consolidated and strengthened the control of insurers over both patients and physicians.

Chapter 23

STRENGTHS AND EFFECTIVENESS OF VETERANS ADMINISTRATION HEALTH CARE SERVICES

In 1992, I moved from the Massachusetts General Hospital to the Boston Veterans Affairs Hospital in Jamaica Plain. When that hospital had been built in 1952, its first chief of surgery, Dr. Henry Faxon—also from MGH—had asked me to serve as a consultant in surgery. This arrangement continued until my transfer as a full-time employee in 1992. In that role, I had active contact with resident surgeons from both Tufts and Boston University Medical Schools (as well as those from Harvard and MGH), which had been very instructive.

Upon transferring to full-time status, I was confronted with the need to use the electronic medical record system at the VA. The transition was not easy! After I had gained some proficiency and realized the superiority of the system as compared to the filing of hand-written paper records, I inquired where the system had been developed. "Here, based on the MUMPS system of Dr. Octo Barnett at the Massachusetts General Hospital!"

"Can you buy it?"

"Yes."

"What does it cost?"

"Just the cost of having a copy made of the software. This was developed using government funds and cannot be copyrighted!"

So I wrote to the administration at MGH, suggesting that they should review this system before investing heavily in another. I had no reply, and I believe my suggestion did not motivate action. Currently, the MGH is engulfed in a (yet another) multimillion dollar restructuring of its clinical data computer programs.

However, I began to see the value of the application of such a system to the population at large, especially the advantage of having a uniform, nonprofit electronic record system for the entire population. I began to advocate to the Massachusetts Medical Society that they should take on this project as a public service as an adjunct to the publication of the *New England Journal of Medicine*. As time progressed and the scope of need enlarged, this was extended to advocacy to include all of New England. Finally, when the total need became more evident, I pushed for the development of a multidisciplinary, nonprofit joint commission (analogous to the joint commission [for hospital inspection] composed of all the major medical professional organizations) for managing a uniform national system of electronic health records for the entire US population. I made a trip to Chicago to meet executives of several main medical professional societies—only to have the executive vice president of the American Medical Association spot me in a hotel corridor then turn and run in the opposite direction! This proposal has been torpedoed by a combination of organizational inertia and entrepreneurial greed. We are now in an extravagant and ineffective period of electronic health record chaos that is likely to have significant longevity and personal cost to American citizens.

As a participant in the Veterans Administration Health Care Services for many years, I have had an opportunity to study its effectiveness. Particularly, since the 1990s, VA health services have steadily increased in quality and effectiveness. This change was catalyzed by Dr. Shukri Khuri, chief of surgery at the Boston/West Roxbury Veterans Administration Hospital and by Dr. Jennifer Daley and Mr. William G. Henderson who developed a system of extracting pertinent, significant, and standardized information from patient care records for the purpose of evaluating the effectiveness of surgical care. Dr. Khuri and many of his associates in the Department of Veterans Affairs designed the National Surgical Quality Improvement Program (NSQIP) for use in 132 VA medical centers which performed major surgery. The effects of this program have been to stimulate progressive improvement in the quality and effectiveness of surgical care which continues today in both VA and civilian hospitals.

In 1999, application of NSQIP in private hospitals began in Emory University (Georgia), the University of Kentucky, and the University of Michigan. In 2001, the (modified) program called ACS NSQIP was undertaken by the American College of Surgeons, and the program has spread broadly across the United States with modifications for hospitals in a variety of settings. The consensus view is that this program has made a significant contribution to the quality of care provided to surgical patients in America.

These strengths of the VA Health Care System, combined with consideration of population health-care needs, have prompted me to consider a modified health-care system suitable for application to the entire United States low-income civilian population. These are the essential features of this proposal include:

1. Development and modular implementation (state by state) of a uniform system of high-quality health care for the low income/indigent population to replace the current Medicaid system.
2. All citizens with incomes below the Medicaid threshold.
3. Governance of this program is proposed to reside in a new structure with the board of directors consisting of a chairman and minority number of members to be appointed by the federal government—the majority of members to be appointed by state governments.
4. State-by-state implementation to allow adjustments in policy as appropriate to eliminate errors or inefficiencies and to prevent gross systems failures.
5. Governing board to have authority to establish goals and budgets and to recommend to Congress a suitable level of funding appropriations
6. Patient services to be modeled upon present VA functions
7. Communications to be based upon the VA system of universal electronic medical records which cannot be patented
8. Staffing of the system should be analogous with the present VA Health Care System—all employees hired and paid by the federal government
9. Staff recruitment should be based on early declaration of vocational intent by nursing and medical students, then financial support through professional training followed after graduation by year-for-year service in the federal Medicaid system at government salaries before being released to practice in civilian status. Currently the VA has affiliations with 107 medical schools, 55 dental schools, and more than 1,200 other health-care schools spanning the entire country. Each year about ninety thousand health professionals are trained in VA medical centers. To meet the staffing needs of an expanded Medicaid system, these resources could be rapidly augmented.
10. The net effect of having new trainees continues; postgraduate training in a structured educational environment would significantly enhance the quality of care for the civilian population as a whole.
11. A major VA contribution to American society is its extensive contribution to medical research. The entire Medicaid population would be brought into this orbit.

Other dynamic forces have intervened in health-care policy, so there has been little consideration of this proposal. However, it is assured that the Veteran's

Health Care System will make additional important contributions to so-called health-care reform. Utilization of a state-by-state implementation would give rapid identification of flawed policies and permit corrections without massive detrimental effects on the entire national public health program.

Chapter 24

CLINICAL RESEARCH

Clinical research refers to that research done in the clinical setting which may have direct effect on our treatment of patients or may extend our knowledge of physiology and pathology. It was always of great interest to me from the time of my medical student experience with congenital methemoglobinemia and the discovery of the catalytic action of methylene blue in reducing this compound to normal hemoglobin. The most amazing aspect of this discovery in January 1943 is that, still, seventy-three years later, methylene blue remains the only effective treatment for methemoglobinemia!

The Cocoanut Grove Fire (on November 28, 1942) with its sequelae of burns, infections, scars, malnutrition, and pulmonary crippling was a huge enterprise in clinical research, including the introduction of penicillin treatment in the United States. At that time, the dose was three thousand units (miniscule by today's standards). Every dose had to be injected by an intern or resident (not nurses), and the patient's urine had to be collected for recycling the penicillin! Those two hospital wards (then designated White 6 and White 10) became *de facto* surgical research wards for shock, burns, blood transfusion, respiratory burns and infections, metabolic changes, nutrition needs, local dressings and debridement and grafting treatment of wounds, and use of portable x-ray units—even psychological changes induce by extreme trauma! Our wards became beehives! But great benefits were derived from those lessons learned, extending forward and continuing today.

When I began working with Dr. Arthur W. Allen, I found him to be constantly involved in clinical research. He and his associates kept track of each of their patients, studied the outcomes of treatment, and constantly worked to devise more effective and less traumatic ways of treatment. His research involved chiefly the healing of wounds, the prevention of thromboembolism, and the gastrointestinal tract, and I was quickly drawn into the arena!

The human body is the most complex machine created in the history of the universe; and the surgeon is often called upon to modify or tweak some malfunctioning part to restore it toward normal. Even at our best (most informed, most skilled) performance, we are always second best to nature's harmony and beauty. The surgeon's mandate is to be a perpetual student and scribe: in constant study of the anatomy, physiology, and function of the body, in recording the findings, and in devising methods of restoration of function toward that beautiful normal state bequeathed by our Creator. I commend to my readers the insight of the great physiologist, Dr. Walter B. Cannon, as expressed in his book *The Wisdom of the Body.*

Clinical research has become an extremely complex enterprise with intensive regulations and institutional and societal control. One of its final phases is the submission of the report for publication in a peer-reviewed journal. The manuscript is reviewed by a panel of knowledgeable experts in the field who submit recommendations to accept or reject the document for publication. New concepts may be at a disadvantage for acceptance in this environment. In my own experience, I have had some ideas rejected that my own work has taught me are valid but are not accepted by such panels. Because I hope these concepts will be pursued further after my own departure, I am going to lay out somewhat dogmatically some results of my own experience that I know to be valid yet have not been generally acknowledged or accepted. In fact, many of these concepts seem not previously to have been recorded.

But before putting forward this list of scientific interest and discovery, I wish to comment on my principal collaborators: Lamar Soutter, MD; Claude E. Welch, MD; and Alfred L. Weber, MD.

Dr. Soutter was one of the pioneer thoracic surgeons, devoted to public service, and a restless innovator in medical education. On D-Day (June 6, 1944), he jumped with the US paratroopers in France. He organized and became founding dean of the new Massachusetts School of Medicine in Worcester, Massachusetts. We worked together on a study of the bacteriology and effectiveness of various methods and antibacterials used in surgeons' hand scrubbing prior to operation, settling on an alcohol-based method that was used by the MGH for over thirty years. In later years, Dr. Soutter asked me to do a required significant surgical procedure on a family member—the ultimate expression of confidence.

Dr. Welch was in the ascending phase of his professional career when I joined the Allen team in 1950. He rapidly became one of the most renowned surgeons in the world. He was a prolific investigator, writer, and teacher, and his name remains a household word in surgery.

Dr. Weber was born on a small farm in Germany on September 23, 1926. He was drafted into the Hitler Youth Movement and, later, the German Army (for World War II). Thereafter, he attended medical school in Germany, then came to the United States, took an internship in New York City, then became a resident in radiology at the Massachusetts General Hospital. He and I met and became fast friends while I was reviewing x-ray studies on the Allen team patients beginning in 1955. This began a lifelong friendship and professional collaboration that was, sadly, terminated by his death in 2014 due to carcinoma of the gallbladder.

Many of these research projects have been individually studied and will be so identified.

Incisions and Wound Healing

Embryological development of the fetus is from the head downward in successive segments called somatomes, wrapping forward from the spine to the anterior midline. As a corollary to this, the anterior midline is the end of the line for all physiological services: cellular nutrition, respiration, and cellular waste disposal. Generally speaking, operative incisions that follow these transverse lines (Langer's lines) inflict less functional disturbance than vertical cuts. "The body loves to be cut and heal crossways!" quoted Dr. Allen. This point has a great deal to do with development of abdominal hernias, which are a huge abomination in modern abdominal surgery, most of which follow vertical incisions. Initially, with Dr. Allen, we used paramedian—never midline—incisions (through the mesial one third of the rectus abdominus fascia with lateral retraction of the rectus muscle belly). Later, left or right lower abdominal oblique incisions (following their description by Dr. Frederick A. Coller of the University of Michigan). After Dr. Allen's death, I extended (in appropriate cases) the left oblique flank incision across the rectus muscles above the symphysis pubis, with division and subsequent meticulous anatomic repair of both the muscles and the fascia. This gave wide abdominal exposure with low retraction tension on tissues at the perimeter of the incision (which minimizes ischemia of the edges of the incision). (For the benefit of the lay reader, pressure at the arterial end of a capillary is 30mm Hg. Greater pressure produced by tissue retraction/tension results in pressure on the capillary, ischemia, and cell death in the wound perimeter).

I used this incision for thirty years with the development of only one incisional hernia (in the upper portion of an incisional wound in the left flank in a slender, elderly woman). These incisions were more comfortable than vertical incisions and healed quickly with good functional restoration. They also gave very satisfactory exposure to the abdominal contents. On one occasion, operating on a mildly obese elderly woman using this incision with one assistant and one nurse, with no mechanical retractors, I did in a single procedure a cholecystectomy, hemigastrectomy with Billroth II anastomosis, bilateral truncal vagotomy, and repair of hiatus hernia. The woman recovered promptly with no complications on long-term follow-up.

Another consideration relates to the bursting tension exerted upon a healing wound. The abdominal wall participates in the muscular effort of every bodily function except those of the eyes and face! It is an established fact that the bursting strength of healing wounds is only about 30% of normal after thirty days and about 80% after three months. Why should we not provide interim elastic support for healing wounds of the abdomen? In my training, I was taught that this practice was improper because it inhibited respiratory exchange. In fact, I observed that elastic support helped respiratory exchange because it relieved some of the pain of respiratory effort in healing abdominal wounds. Throughout my career, I employed this strategy.

To summarize my experience and my observation of the experience of others, in managing extensive abdominal surgery, the abdominal wall itself is a vital organ requiring meticulous attention in opening and closing incisions. The abdominal wall may be endowed genetically with the capacity for fairly rapid stretching and recovery to accommodate space requirements of pregnancy. Specifically, I prefer oblique or transverse incisions to minimize injury to nerves and blood vessels, sharp dissection over cautery, avoidance of tissue injury by prolonged mechanical retraction, figure-of-eight or continuous sutures for fascia repair to minimize tissue tension through invocation of the pulley principle (tension on a pulley rope (string) varies inversely as the number of strands). I advocate "horse-hair" drains of multi-strand suture bundles just above the deep fascia closure to provide drainage of wound serum by capillary attraction, closure of Scarpa's fascia with continuous, fine, absorbable, monofilament suture, and skin closure with subcuticular, continuous, fine absorbable, monofilament sutures for all types of wounds, including clean, clean-contaminated, and contaminated. Exteriorization of all extravascular wound fluid collections enables the body's native defenses to operate efficiently: capillary blood flow within 3–5 cell widths (20 μm) is the essence of cellular (and all!) life. Despite all our best efforts at aseptic technique, bacteria are still omnipresent and are only repelled by the animate defense of the body.

The Alimentary Tract

In the mid-twentieth century, carcinoma of the stomach was very prevalent in America. During my surgical training, partial or total gastrectomy for this malady was among the commonest of abdominal operations. There were also frequent gastric operations required for the treatment of benign duodenal and gastric ulcers. Humans are omnivores; that is to say, we are capable of digesting proteins, fats, and carbohydrates. Our own tissues are mainly proteins and fats, so we have the capacity, under certain circumstances, to digest our own tissues. A large proportion of these tissue-digesting enzymes are secreted at the upper end of the digestive stream: the mouth, stomach, liver, and pancreas. These ferments come together for a concentrated attack on our food (and, sometimes, ourselves) in the duodenum just beyond the stomach. It may not be a pretty picture, but without that conjunction, we should not survive!

Definitive gastric surgery began in 1881 at the University of Vienna when Theodor Billroth did the first gastric resection for carcinoma of the stomach. His first patient died—a portent for the difficulty and danger of surgical intervention in the nidus of tissue digestion. However, none of the subtleties of digestive physiology were then known. There followed a century of physiological and surgical study to establish safe and effective surgical management of these dominant gastrointestinal conditions: carcinoma of the stomach and peptic ulceration of the stomach and duodenum. A succinct summary and bibliography of these studies may be found in "Safe Management of the Impossible Duodenum" (Rodkey, G. Arch. Surg. 123:558–562, 1998).

However, in 1982, the Australian scientists Barry Marshall and Robin Warren discovered that the bacterium *Helicobacter pylori* causes most instances of benign gastric and duodenal ulcers and may cause gastric cancer. Treatment of this bacterium by various combinations of antibiotics and proton pump inhibitors has greatly reduced the incidence of peptic ulcerations of both duodenum and stomach and has sharply diminished the role of surgery in management of this disease.

The Post-Gastric Digestive Melee

All our ingested food breaks down into fats, carbohydrates, and protein. The function of digestion is to prepare these ingredients to be absorbed into the blood stream to be transported for further modification (liver) or utilization in cellular metabolism. In general terms, carbohydrates are largely converted to glucose and absorbed through the intestinal mucosa into the blood stream thence to the liver. Fats are emulsified in the gut lumen by the action of bile and absorbed as fatty acids through lymphatics in the gut wall and, via the thoracic duct, pass directly into the systemic venous system.

Protein digestion is more complex and results in conversion of protein foodstuff from complex compounds to amino acids, which are then absorbed into intestinal capillaries and—through the portal venous system—transported to the liver. This process begins in the stomach with pepsin and pepsinogen and continues in the duodenum and small bowel with addition of trypsin and chymotrypsin from the pancreas. Amino acids are thence transported through the portal venous system to the liver.

It is well-known that the fastest cellular regeneration in the human body occurs in the mucosa of the duodenum and upper jejunum, and it may be reasonably conjectured that this was nature's method of compensating for rapid cellular turnover from autodigestion of tissue here by concentrated trypsin.

The pancreatic and common bile ducts join for a short distance (1–1.5 cm) in their oblique passage through the wall of the second portion of the duodenum. This common segment is surrounded by an encasement of smooth muscle which is called the sphincter of Oddi. The common wall between the terminal portion of each duct is a thin membrane of tissue which acts as a "flap valve," tending to shut off reflux to the other side when flow is generated by pancreatic secretion (up to 250 mm saline secretory pressure); or conversely, when flow is generated by pressure of hepatic secretion of bile (up to 190 mm saline) or due to gallbladder contraction (??, but > than 250 mm saline). It is worth noting that the gallbladder is the pump generating the highest pressure in this hydrodynamic system.

Trypsin has an optimum activation pH range from 7.5–8.5, so it is not actively proteolytic as it is secreted by the pancreas, and it is not immediately activated by admixture with bile; however, if pancreatic secretion is incubated with bile at 37 degrees centigrade for twenty-four hours, it becomes highly proteolytic. Thus, when—as regularly happens in many individuals—between meals and during sleep, if the pancreas is actively secreting due to vagal nerve stimulation and the sphincter of Oddi is closed, both bile and pancreatic secretion as a mixture reflux retrograde into the gallbladder, which is concentrating its contents by absorption directly into the liver by a factor of fivefold to eightfold.

Many years ago, the physiologist Mann demonstrated that the injection of a bile/pancreatic juice mixture into the pancreatic duct does not cause pancreatitis. However, if bile and pancreatic juice are incubated together at 37 degrees centigrade for twenty-four hours then injected into the main pancreatic duct, a roaring acute pancreatitis ensues. (Mann, F. C., Giordano, A. S. The Bile Factor in Pancreatitis. Arch. Surg., 6:1–30, 1926)

Retrograde flow of pancreatic secretion into the gallbladder is a regular, normal phenomenon and can be reliably detected by testing gallbladder bile for the presence of amylase. If the bile/blood amylase concentration ratio is >1, the excess over 1/1 is always due to amylase of pancreatic origin. Various conditions (including fasting) may delay the emptying of the gallbladder and cause some activation of pepsin proteolytic activity within that organ (incubation at 37 degrees centigrade). Gallstones, when present, are *always* found to be either solitary or in colonies of stones—each colony having a distinct similar size of individual cohort stones. This means that stones are precipitated in episodic events—most plausibly, episodes of enzymatic digestive cellular injury to the gallbladder mucosa secondary to stasis incubation of bile and pancreatic juice in the gallbladder lumen. Precipitation of stones occurs when necrotic mucosal fragments fall into the supersaturated colloidal suspension of fluid which is bile. This observation, of critical importance, has never before been documented. I believe.

The same sequence of activation of trypsin in the gallbladder by stasis/pancreatic juice incubation (secondary to fasting, lack of cholecystokinin secretion) is almost certainly the cause of acute cholecystitis, acute gangrenous cholecystitis (with or without gallstones), cancer of the gallbladder, chronic cholangitis, and cholangiocarcinoma. It is highly significant—and consistent with this theorem—that 95% of adenocarcinomas of the pancreas occur in the head of that organ, in its ductal system, precisely where the incubated bile-pancreatic juice mixture is refluxed during gallbladder contraction.

There are many reasons to suspect that repetitive episodes of mucosal autodigestion, inflammation, repair, hyperplasia, metaplasia, and neoplasia cycles are related to Crohn's disease and chronic ulcerative colitis, as well as benign and malignant tumors of the entire enteric tract and its embryonic branches. This thesis is the most important avenue for future investigation. For example, over the course of the last sixty-five years, it has become established that the regular use of aspirin decreases the expected incidence of carcinoma of the colon. How does this happen? Aspirin is, chemically, acetylsalicylic acid. That is, aspirin is a substance that reduces the pH of the liquid stream within the lumen of the gut. Could this chemical phenomenon decrease the proteolytic (autodigestive) activity of trypsin within the gut lumen and inhibit the mucosal inflammation, digestion, hyperplasia, metaplasia, neoplasia cycle? An observation which supports this thesis is that vegetarians have significantly lower rates of trypsin digestion and of bowel polyps and cancers than control cohorts eating more protein in their diets.

This needs experimental investigation. Conceivably, release into the lumen of the small bowel of a nontoxic acid substrate from an enteric coated capsule (enteric coated aspirin?) might enhance this reaction and significantly reduce intestinal inflammation, affecting Crohn's disease, chronic ulcerative colitis, and polyp and tumor formation in the intestine.

The Peritoneum/Abdominal Cavity

In the 1950s, surgeons became aware that talcum powder, then used on rubber gloves to facilitate their donning and removal, was causing inflammatory changes within the peritoneal cavity. In response to this observation, other less irritating substitutes were found for this purpose. However, during this period, I took many specimens of adhesions encountered during operations and sent them for histological examination in the pathology laboratory of the Massachusetts General Hospital. To our great surprise, although some talc granulomas were found, the adhesions far more frequently encased textile fibers—cotton staples. These are tiny filaments of up to 5/8 inch in length and are the lint shed during the lifetime of usage of every cotton fabric. Were these filaments causing adhesions secondary to foreign body-induced inflammation?

In the early 1970s, I became the MGH representative to the Ethicon General Surgery Advisory Panel. I asked the company to manufacture a surgical sponge that did not contain cotton. Within two years, they created Sof-Wick—sponges manufactured with cellulose fibers that did not shed lint. I began using them at once and discarded all cotton fabrics from my operative fields. In addition, I educated all my fellow surgeons at the MGH regarding the probable superiority of this practice. Of them, only one—Dr. Ronald Malt—joined me in the exclusive use of Sof-Wick sponges and pads within the abdominal cavity. During the next twenty years, I noticed a marked reduction—essentially complete absence— of intraperitoneal adhesions upon secondary intra-abdominal operations (as in deliberately staged procedures). In addition, I found that patients who had abdominal operations prior to mine (presumably with use of cotton sponges) developed more postoperative abdominal pain, fever, and ileus than those who had a virgin abdomen. I interpreted that phenomenon to reflect the release of previously encapsulated cotton staples to reignite the inflammatory cycle.

Unfortunately, Dr. Malt ceased operating in the late 1980s, and I moved from MGH to the Veterans Administration Hospital in Boston in 1992. After this, MGH stopped buying Sof-Wick sponges and Ethicon sold the patent. I was not able to get the VA to conduct a trial of Sof-Wick surgical sponges. The cost, disability, morbidity, and mortality of complications of adhesions due to cotton staples certainly merits further experimental investigation of alternative textiles to eliminate cotton from the operating theater—a condition which I consider to be a professional imperative.

Pilonidal Sinus

The term "pilonidal" means, literally "hair nest" and is applied to abscess, sinus formation or inflammation in the intergluteal crease over the lower sacrum and coccyx. During World War II, this was popularly called jeep disease because of its prevalence among soldiers who frequently rode in those military vehicles. In the mid-twentieth century, the cause of this disease was obscure, and a congenital defect was commonly suspected. However, many observations are detracted from this hypothesis. Pilonidal sinus rarely occurs in women, and then, only in hirsute women. Pilonidal sinus never occurs among Chinese—who have little, if any, body hair. Pilonidal sinus occurs between the fingers of barbers. It may occur in the axilla or in the umbilicus. Histological studies show that there are *never* hair follicles in a pilonidal sinus, only hair shafts. Microscopically, hairs in a pilonidal sinus look like little porcupine quills: sharp-pointed, the shaft flanged with tiny spicules to keep driving the point forward. Customary treatment was excision of the inflamed area, including the contained hair, followed by local dressings until the wound healed by slow stages.

My own epiphany came when a new patient—a big, red-headed, hirsute fireman—was sent to me for treatment of a recurrent pilonidal sinus with abscess, pain, and bleeding. This individual had five previous operations for the same problem, and when I examined him, I was shocked to find that the floor of his wound was the posterior wall of the rectum! No further excision was possible! I cleaned his wound, shaved all the hair from his waist down to midthighs, and put him on a regimen of wound dressings and once-weekly application of depilatory cream over the skin from his waist to midthighs. To my amazement, the wound rapidly healed and never recurred!

I realized that the cause of the malady was migration of shed hair shafts into the moist, soft intergluteal crease where the little "porcupine quills" bored through the skin to set up housekeeping, bringing with them infection and inflammation.

Thereafter, based on the principles observed in this case, I devised a simple operation, performed with local anesthesia, to be done in the office and permit the patient to go home. This was a midline incision over the abscess, excision of necrotic tissue and removal of all the hair, placement of a multistrand suture drain in the base of the wound, and wound closure by subcutaneous approximation to presacral fascia without tight juncture of the skin edges. This permits rapid wound healing which must be preserved by regular application of depilatory cream from the waist to the midthighs *in perpetuity*!

Remarkably, the world surgical literature still publishes many articles describing arcane (and often quite deforming) surgical treatments for this problem—frequently including rotating gluteal muscle flaps to fill large presacral excavations.

I shudder!

The human gluteus maximus is our primary muscle of rest and repose. We spend more time at work and at play sitting on those cushions than in any other posture! Further, obliteration of the intergluteal cleft with subsequent abnormal tissue tension causes permanent discomfort.

Primum non nocere!

The Gut
Sustainer of Life

Digestion is the process of converting meat, vegetables, grains, fruit, and other edibles composed of proteins, fat, carbohydrates, and other essential minerals, vitamins and water into an absorbable form to be processed internally by the body. This is accomplished by a combination of mastication, muscular churning of the gut, and chemical breakdown by enzymes (organic catalysts) to convert raw foodstuffs to amino acids and glucose (absorbed through the intestinal mucosa into the intestinal capillaries, veins, and the liver) and fatty acids to be absorbed through the intestinal lymphatics and transported directly into the blood via the thoracic duct. In addition, digestion is aided by countless trillions of bacteria in the gut—most unknown and unnamed. And the entire digestive system is fortified by specialized cells in the mucosal lining to provide immunological protection from bacterial, virus, or fungal invaders. All this activity functions under the unconscious control of the autonomic nervous system, hormonal control, and cell-to-cell communication, which frees the owner to read books and watch baseball games! Scant wonder that the gut has an occasional burp that summons the surgeon!

The organs of digestion (beginning with the salivary glands) reside chiefly within the peritoneal cavity of the abdomen. This cavity, surrounded by a muscular wall that we have previously described, is lined by a delicate, cellular endothelial layer called the peritoneum, which secretes a thin layer of lubricant fluid and which has the capacity to provide rapid sealing, isolation, and repair of injuries— traumatic, enzymatic, or infectious. This layer also covers the external surface of all the abdominal viscera: stomach, small and large intestine, urinary bladder, the uterus, fallopian tubes and ovaries, liver, pancreas, kidneys, spleen, and omentum. The undersurface of the diaphragm has extensive fields of small pores that open into lymphatic channels with microscopic flap valves propelling free peritoneal contents through the thoracic duct back to empty into the venous system at the base of the neck. This system is constantly pumped by diaphragmatic (respiratory) and abdominal wall movements.

The omentum is a fatty, lace-like veil suspended from the stomach and colon, lying between the intestine and the anterior abdominal wall. It's the first responder to all abdominal catastrophes to provide containment, cleanup, and healing resources—as well as being a repository for production of hormones, immunologic, and stem cell functions. Altogether, the omentum is still a poorly understood and undervalued benefactor to its owner! In my early career, I knew one famous surgeon who cut out and threw away the omentum (as "being in the way and likely to cause trouble") in every case in which he opened the abdomen, reminiscent of others who frequently cut out the umbilicus (belly button) as "being in the way"! Every human has an umbilicus. How isolated do you become without one? Every insult to the human body is translated to a violation of the self—whether mechanical, chemical, or psychological or all. And in so far as possible, it's to be avoided. Any bodily structure which has survived millions of years of genetic fine-tuning has a survival function—whether or not you may happen to know what it is!

I am thinking of the incidental appendectomy! When carcinoembryonic antigen (CEA) was identified in colon cancer tissue extract by Drs. Phil Gold and Samuel O. Freedman (1965), I thought to myself, *Who is to say that the appendix may not exert lifelong surveillance to suppress cells that produce this—or some other unidentified biological messengers which cause malignant cellular change?* The appendix is not a digestive organ; it may be an endocrine-primitive cell organ. The thought occurred to me: *Any organ which has survived millions of years of genetic fine-tuning has a function—whether or not you know what it is!* On that day, I stopped doing incidental appendectomies,—which was the usual custom then (and, perhaps, still). There have been subsequent studies purporting to show increased incidence of colon cancer among patients who have had appendectomy.

The process of food digestion is an extremely fine-tuned, harmonious interaction among muscular propulsion, nerve (autonomic) control, glandular secretions of specialized chemicals and enzymes to break down fats, proteins, and carbohydrates into absorbable moieties, aided by bacterial interaction in totally unknown ways. We do know that protein digestion has potential hazard for the host for we, too, are protein. This is an area of physiology that is only at the threshold of understanding and that has been a consuming interest of mine for the past fifty years. It used to be said, "You become what you eat." Now we ask "Do we really eat ourselves too?" As yet, unresolved. Although it is not yet scientifically accepted, I believe that chronic autodigestion (chiefly by trypsin) with secondary inflammation and repair is the root cause for acute and chronic ulcerations of the gastrointestinal tract, for acute pancreatitis and pancreatic cancer, for gallstones, chronic and acute cholecystitis, for adenocarcinoma of the gallbladder and the biliary ductal system including cholangiocarcinoma, for polyps of the small and large bowel and adenocarcinomas of the same distribution, and for chronic enteritis (Crohn's disease) and chronic ulcerative colitis. Undoubtedly, there are genetic differences among individuals with varying susceptibility to these diseases still to be defined.

The summation of this section is, we know so very little about the excruciatingly complex functioning of the human body, even ignoring completely our governing integrator—the brain! We take it all for granted, we do not protect it from physical or chemical abuse, and we are not grateful for the enormous blessings of everyday activities. When our scientists glimpse a new physiologic function or pathway, we celebrate by awarding a Nobel Prize—never realizing that the function has been operative for uncounted aeons before our new Nobelist arrived on the scene! St. Paul, paraphrasing Jesus Christ, expressed it most eloquently: "Know ye not that your body is a temple of the Holy Spirit, which is in you, which you have from God, and you are not your own for you were bought with a price: Glorify God, therefore in your body" (I Corinthians 6:19–20). A physician and, especially, a surgeon function at the interface of all these confusing concepts.

In this discussion, we have ignored the brain. In the old days, surgeons passed down a warning: If you have a patient who tells you before the operation, "I am going to die!" don't operate on him! He may, in fact, be in grave danger of dying during your operation!

Although I have never seen any scientific studies of the issue, during this past century, our growing understanding of cardiac response to adrenergic hormones is compatible with this "old wives' tale." For this reason, never in my entire career did I schedule an elective operation on a Friday the thirteenth. It is inevitable that any person undergoing anesthesia and operation (total surrender of self-awareness and control) should be thinking, "Will I survive this?" What increment to this anxiety will be added by Friday the thirteenth? Whatever it is, I didn't want it!

Of course, this observance has stimulated curiosity as to the origin of the superstition. I have never been able to find any discussion of this issue (although there must be some in the world's literature) nor did any people whom I have asked have any insight regarding the matter. In my own thinking, however, the origin relates to the Last Supper, with the Lord and his twelve disciples (thirteen souls and a Friday). And in these ruminations, I have come to the realization that the "Holy Grail" was the cup containing the wine at that Supper.

More Issues Related to Research

In 1970, together with Dr. W. Gerald Austen and Dr. Alfred L. Weber, I submitted an article to the *New England Journal of Medicine that w*as summarily rejected. However, I know that the material is factual and that the conclusions are valid. Further, I had a very intensive bibliography which outlined the physiologic bases for my studies. Essentially all these studies were done before establishment of PubMed in the 1950s and, thus, are never accessed by modern bibliographic searches. For this reason, I shall also include the bibliographic references.

Pressure-Flow Relationships in
the Pancreatic-Biliary System
Grant V. Rodkey, MD
Gerald Austen, MD

Alfred L. Weber, MD

The possibility of anatomical confluence of the common bile duct and the duct of Wirsung at the ampulla of Vater has been established since Oddi's report in 1887 (132). Mechanical conversion of the system into a common channel with reflux of bile into the pancreatic ductal system was reported by Opie in a case of acute hemorrhagic pancreatitis in 1901 (137). Archibald suggested that spasm of the sphincter of Oddi may produce temporary obstruction and divert bile into the pancreatic duct. He performed sphincterotomy in dogs, demonstrating reduced flow resistance from 500 mm of water to 100 mm of water (5). Subsequently, many studies of bile reflux into the pancreatic duct have been carried out, with somewhat inconclusive results in explaining the mechanism of pancreatitis.

Less generally known, but firmly established, is the fact that pancreatic juice may reflux into the biliary system as well. Wittich in 1872 noted amylase activity in drainage from a cholecystocutaneous fistula and interpreted this as evidence that pancreatic juice was transmitted through the gallbladder (194). The presence of amylase activity in gallbladder bile was confirmed by Westphal, who also found pancreatic enzymes in common bile duct drainage (185). He believed these enzymes are capable of causing acute and chronic cholecystitis, cholangiolitis, and biliary cirrhosis (186). Popper studied amylase in gallbladder bile of 200 patients with assorted biliary, pancreatic, and gastrointestinal disorders. He found elevated amylase levels in 20 of these—of whom 17 had gallstones. One had a normal biliary tract with associated gastric ulcer and two had mild chronic cholecystitis. One of the latter cases had serum amylase of 4, bile amylase of 32,000 Wohlgemuth units. Culture of the bile in this instance showed no growth. It was his conclusion that pancreatic enzymes in the gallbladder might cause pathological changes if there were obstruction in the biliary tract (144). Popper considered this to be the mechanism of bile peritonitis without gallbladder perforation in which he found amylase levels of 1024 Wohlgemuth units in the gallbladder, 4096 units in the peritoneal fluid, and 32 units in the blood (143). This observation complemented the earlier studies by Blad who used fresh gallbladders as dialysis chambers and demonstrated that addition of pancreatic juice to bile in the gallbladder resulted in passage of bile salts through the intact gallbladder wall. Blad attributed this phenomenon to the enzymatic digestion of colloid components of the bile, thus freeing the smaller bile salt molecule which became dialyzable (16). Colp and his coworkers observed three cases of acute cholecystitis with pancreatic enzymes in the gallbladder bile. In two of these cases, bile-stained fluid was present in the peritoneal cavity, and culture showed the bile to be sterile (30). Golf and Doubilet reported the finding of large quantities of amylase in the gallbladders of four patients with acute cholecystitis. They found amylase in bile drained from the common bile duct through a T-tube in 8 cases and noted one case which drained clear pancreatic juice through the T-tube (29).

Anatomical studies have confirmed the potential for common channel formation between pancreatic and biliary systems in the majority of human subjects. Thus, collected autopsy studies by Reinhoff and Pickrell indicated a common channel in 85% of 476 cases (155). In a similar study, Hansen collected 2,426 cases with approximately 70% incidence of anatomical common channel (64). Using cholangiographic techniques in 130 postmortem cases, Milbourn demonstrated reflux into the pancreatic duct in 91.5% (124). Hjorth occluded the tip of the papilla of Vater with a clamp and injected the main pancreatic duct in postmortem studies in 50 males and 50 females. He found that the common bile and pancreatic ducts joined at the base of the ampulla of Vater in 86%, the pancreatic duct entered the ampulla near its termination in 6%, and the pancreatic duct entered the duodenum through a separate opening in 8% (72). With a 3 mm stone impacted at the ampulla, fluid introduced into the hepatic duct at a pressure of 100 mm of water has been shown to enter the pancreatic duct in 66% of 100 fresh necropsy cases.

Anatomic dissection in this series demonstrated an anatomic communication of the biliary and pancreatic ducts in 74% (23).

In the living subject, these relationships are modified by the tonus of the sphincter of Oddi—the anatomy of which has been elucidated by Boyd (18, 19). Simultaneous cinefluoro-cholangiograms with recordings of action potentials from paired electrodes implanted into the sphincter of Oddi and another pair implanted into the duodenal wall in man have confirmed Oddi's and Boyden's theory that relaxation of the sphincter permits bile flow, and that this muscular activity is independent of the contraction of the duodenal wall (134, 135, 136). Since this relaxation of the sphincter, body proceeds from its proximal end toward the papilla and contraction is in reverse order, only those instances in which the ducts join near the base of the ampulla are very likely to be converted into a physiologic common channel, as is confirmed by angiogeographic studies (24, 82, 131). Various series of operative and postoperative angiograms in patients have demonstrated reflux into the pancreatic duct in from 25% to 45% of cases (11, 73, 79, 95, 131, 162).

Factors that control the tonus of the sphincter body include pharmacological agents, neural, and hormonal influences in a complex and interrelated pattern. A summary of present knowledge of these control mechanisms is listed in table I. Normal resistance of the sphincter is in the range of 150 mm of water, and it may rise to 400 mm or higher under the influence of those factors promoting spasm, or fall to the range of 70 mm within twenty minutes after eating a fatty meal (12, 134).

Pressure within the common bile duct is also influenced by peristaltic activity and muscular tone in the duodenum (38, 92, 147, 176, 181). However, the oblique passage of the common bile duct and pancreatic ducts through the wall of the duodenum usually serves to protect against reflux duodenal content even the presence of bursting pressures within the duodenal lumen (37, 175). Occasional exceptions to this occur, however, and when reflux of duodenal contents into the papilla occurs, it may initiate enzymatic digestion within the pancreatic and biliary systems (103, 104). Vomiting may produce pressures in the range of 120 cm of water in the pancreatic duct without duodenal regurgitation. For all practical purposes, when the intraluminal gut pressure exceeds 400 mm of water, the biliary and pancreatic ductal systems are obstructed (22).

Common duct pressure curves generally fall into three distinct patterns. These are associated with the following events:

1. Contraction and relaxation of the sphincter of Oddi (3–4 times/minute)
2. Rhythmic contractions of the gallbladder
3. Respiratory movements of the diaphragm transmitted by the liver to the biliary system (172)

Large rises in the pressure within the biliary system may accompany sneezing, coughing, and muscular exertion, but since these are transmitted generally throughout the abdomen, they have little or no influence on flow distribution (131).

Pressure and flow relationships in the biliary-pancreatic ductal system are also responsive to the secretory volume and pressure of bile and pancreatic juice. In man, bile secretion varies between 1000–2000 mL per twenty-four hours, and its volume may be increased by administration of cholagogues or decreased by anticholinergic drugs (36). There is a normal diurnal variation in the rate of bile secretion, the largest volume being produced during the day and the smaller at night (5). Resting pressures in the common bile duct of man are in the range of 100–150 mm of water (12, 36, 81, 92). For a summary of some of the factors affecting biliary-pancreatic secretion, please see table II.

Maximum secretory pressures of the human liver are difficult to obtain. In cats, dogs, goats, and monkeys, these average about 300 mL of water (68, 89, 125, 141). Ivy has determined human bile secretory pressure to be in the range of 210–270 mm of bile (78).

The question whether the common bile duct in man is able to contract actively is unsettled. In dogs and monkeys, the common duct apparently has the capacity for intrinsic muscular activity (102, 184). Histologic studies of human ducts have been interpreted to demonstrate the presence of smooth muscle and nerve fibers (32, 66). On the other hand, cine-cholangiograms appear to show peristaltic activity limited to the ampullary region with the duct itself behaving as an elastic tube responding passively to increased pressure by dilatation and elongation (11, 127, 131). In this respect, this behavior seems similar to that of the aorta. Cholangitis damps this mechanism and makes the duct function more like a simple conduit (147).

The 24-hour volume of pancreatic juice in man is quite variable and may range from 700–3300 mL (50, 168). There is a normal diurnal variation with a smaller amount of night secretion which contains maximum concentration of enzymes and chlorides, low volume of bicarbonate (50). In the pancreatic juice, concentrations of the various enzymes apparently always follow a parallel pattern except when influenced by a deliberately imbalanced diet (1, 36, 62, 76, 161, 170). Recent studies by Demand and Kuckuck suggest that exocrine pancreatic secretion is a mixture derived from the acinar cells (chloride, potassium, phosphate, enzymes, and sodium) and from tubular cell secretion (sodium and bicarbonate) (34). The healthy human pancreas with maximum secretin stimulation can produce juice at the rate of 2 mL per kg/hr (containing 90–150 mEq per liter of bicarbonate) (4).

Secretory pressure of the pancreas is somewhat higher than that of the liver, with maximum pressures in dogs, cats, and monkeys ranging from 300–500 mm of water (22, 68, 117). In a particular animal, pancreatic ductal pressures exceed common bile duct pressure in both the fasting and feeding states with occasional brief reversal in the early phase after meals—a phenomenon probably associated with contraction of the gallbladder (3, 959, 120, 122, 139). Measurements in humans have indicated pancreatic ductal pressures in the range of 350 mL of water with common bile duct pressures in the range of 100–200 mm of water. Stimulation with food and with morphine in these cases cause increases in pancreatic ductal pressure of 90 mm of water and in the common bile duct pressure pressures rose only 60 mm of water (176).

The gallbladder functions as the chief motor regulator of biliary ductal flow and pressure, but this role is complex and subtle. Its filling is dependent upon the tonic contraction of the sphincter of Oddi, and species lacking a gallbladder are apparently also without an effective sphincter mechanism at the outlet of the common bile duct (52, 116, 164). There are spontaneous rhythmic contractions of the gallbladder occurring at the rate of 2–4 times per minute with pressure fluctuating in the range of 50–75 mm of water. Emptying of the gallbladder occurs in response to cholecystokinin, which increases amplitude of the gallbladder contractions to develop pressures as high as 200–300 mm of water. Fear, anxiety, or pain may delay emptying of the gallbladder—perhaps by inducing spasm of the sphincter of Oddi (31, 74, 78, 80, 191). The valves of Heister do not impede bile flow in either direction. These structures occur only in animals that have erect posture and are apparently an adaptation to prevent distention or collapse of the cystic duct with sudden pressures or torsional or postural changes (98).

Emptying of the normal gallbladder is usually not complete, and after sectioning of the left vagus nerve trunk or the hepatic plexus of the anterior (left) vagus nerve, its emptying is even less efficient. Moreover, after vagus interruption, the resting volume of the gallbladder increases to about twice normal. This is interpreted as evidence that vagal impulses maintain the muscular tonus of the normal gallbladder (44, 159, 173). In vitro studies of the contractility of the human, gallbladder confirms its marked response to cholecystokinin, moderate response to acetylcholine and slight response to secretin (109). The question whether magnesium sulfate applied to the papilla of Vater also causes some contraction of the gallbladder is unsettled (187).

In addition to rhythmic filling and emptying, the gallbladder influences pressure and flow in the biliary system by its ability to concentrate bile. Gallbladder bile may be concentrated five to nine times as compared to the hepatic bile, and absorption through the gallbladder mucosa-hepatic interface takes place at a rapid rate. Very little bilirubin, bile salts, or cholesterol are absorbed, and cholesterol content of the gallbladder bile may be augmented by desquamation of mucosal cells. Inflammatory changes in the gallbladder wall diminish its concentrating ability and may contribute to acidification of gallbladder bile (8, 17, 78,153).

Dynamics of the biliary tract are somewhat altered following cholecystectomy. There may be some absorption by the mucosa of the common bile duct but little information is available to settle this point (49, 158, 188). Postcholecystectomy physiologic dilatation of the duct has been reported in both animals and men (14, 88, 133, 167). However, recent studies using intravenous cholangiograms for long-term observations have indicated that there is *no* significant normal dilatation of the common bile ducts of man following cholecystectomy (101, 113, 114, 151).

A final significant component to the fluid dynamics of the combined biliary-pancreatic system is a damping effect secondary to what may be termed the sump or swamp function of the pancreas. This organ has a remarkable capacity to permit interstitial dispersal of its entire exocrine output with development of only minor interstitial edema and ductal dilatation (70). These secretions are returned to the systemic circulation by the pancreatic (portal) veins as the first and major defense mechanism, but later the lymphatics contribute substantially by return through the thoracic duct (46, 145). Experimentally, Dunphy and Robinson illustrated this function of the pancreas when they produced partial common duct obstruction distal to the entry of the pancreatic duct in goats. This produced large biliary reflux into the pancreas and caused transient elevation of amylase levels but no acute hemorrhagic pancreatitis (45). Doubilet provided an instructive example of this function in clinical cases which he studied by pancreatography under conditions of complete pancreatic duct obstruction. Within five minutes after intravenous injection of 100 units of secretin, pancreatic ductal pressure rose from 70–400 mm of water with gradual return to resting level over the course of the next hour. During the same period, serum amylase levels rose from 150–1500 Somogyi units (39). The degree to which this damping mechanism is involved under physiological conditions is not known, but its potential effect upon pressure-flow relationships in the ductal system is considerable.

Clinical Study

A group of fifty-five patients has been studied with special care to evaluate their potential for biliary-pancreatic reflux, to correlate these findings with pathological changes in the gallbladder and pancreas, and to relate these changes to the role of infection in the biliary tract. Methods used have included sampling venous blood and gallbladder and common bile duct bile for amylase content, aerobic and anaerobic cultures of the bile, operative cholangiography combined with manometric readings, and, in cases of common bile duct exploration, T-tube cholangiograms. Among the group, there were no deaths and no serious postoperative complications. Duration of postoperative hospital stay averaged 9.7 days. Three of the cases merited special consideration and are presented in some detail.

Case No. 1. MR, No. 690860

The patient was a woman, aged fifty-six, who had gallstones noted as an incidental finding at hysterectomy eight years previously. In the interim, she had mild symptoms of fatty food intolerance and occasional epigastric discomfort. Two days prior to admission, she developed severe upper abdominal pain radiating across both costal margins and through the back. She became febrile, developed nausea and vomiting, was unrelieved by opiates at home, and was subsequently admitted to the hospital. On admission, she was acutely ill, dehydrated, and complaining of severe upper abdominal pain radiating to the back. There was marked local tenderness in the epigastrium and right upper quadrant, some rebound tenderness in the lower abdomen. Fever was 102°, leukocyte count was 15,200, hematocrit 44%, and serum amylase 22 Russell units. Following a few hours of observation and rehydration, the patient was explored.

At operation, a small amount of serous fluid was present in the upper abdomen. Gallbladder was distended, edematous, and acutely inflamed. Pancreas was 4 cm in thickness, nodular, and surrounded by some areas of fat necrosis. Antrum of the stomach and the greater omentum were edematous. Gastrohepatic ligament, common bile duct, and hilum of the liver were acutely inflamed and so edematous that it was not possible to dissect the area for careful inspection of the common duct.

The gallbladder was opened, four faceted black stones (each measuring 1 cm in diameter) were removed, and cholecystostomy was performed with a number 14 French catheter. None of the stones was impacted in the ampulla of the gallbladder.

Postoperatively, the patient did well, and on the eighteenth day, a cine-fluoroscopic cholangiogram was performed. The contrast medium was injected by means of a Harvard pump at a rate of 7.64 mL per minute, and pressure was monitored by manometer in the line. Bile and blood amylase and lipase values were monitored before and following the examination.

No residual stones were evident in the entire biliary system. Resting pressures of 156 mm of saline gradually increased to 360 mm over the course of three minutes (23 cm^3) at which time the infusion was stopped. Pressure returned to 182 mm over the next seven minutes.

During this infusion, the patient began to experience upper abdominal pain with radiation across the costal margins and into the back as the pressure reached 280 mm. The common bile duct began to exhibit peristaltic contractions of progressive amplitude and increasing frequency, with radiation of pain to left scapular mesial border. At 284 mm, contrast medium began to reflux into the pancreatic duct. Some residual remained in the pancreatic duct until the pressure had fallen to 266 mm and pain subsided after stopping the infusion (fig. 1).

Following examination, the patient experienced moderate upper abdominal discomfort for a few hours, but by the following morning, she seemed completely recovered. As indicated by the sharp rise in blood amylase and lipase, examination had produced a recrudescence of her pancreatitis (table III). Cholecystostomy tube was removed, and the sinus healed promptly.

Approximately four months after initial exploration, the patient was readmitted for elective cholecystectomy, having been quite comfortable during the interim. At reoperation, the entire subhepatic space was obliterated by adhesions, and there were still some residual areas of resolving fat necrosis. Liver appeared normal. The gallbladder was quite thickened but contained no stones; common bile duct appeared normal. Remarkably, the pancreas also was normal to inspection and palpation.

Gallbladder was removed, and choledochostomy was accomplished. There were no stones in the duct, but the ampulla of Vater was stenotic and would not permit passage of a 3 mm probe. Transduodenal exploration and inspection of the ampulla confirmed a pinpoint opening in the papilla, which was biopsied. Sphincterotomy was accomplished, and the opening of Wirsung's duct was located about 3 mm above the lip of, but within the ampulla of Vater. A 2 mm probe fell into the pancreatic duct for distance of 8 cm with no resistance—a finding consistent with chronic dilatation secondary to stenosis of the sphincter.

Postoperatively, the patient did well and left the hospital on her seventh postoperative day. A repeat cinofluoroscopic examination was done through the T tube. Conditions of the examination were similar to those at the first study:

Resting pressure in the duct was 160 mm saline. Infusion was begun at 7.64 mL per minute. At five minutes (38.2 cc), pressure had risen gradually to 320 mm, and she began to experience mild epigastric pain. Infusion was continued, and at seven minutes (52.2 mL), pressure reached 360 mm. There was no reflux into the pancreatic duct. With the infusion continuing, pressure then stabilized at 350 mm with moderate epigastric pain tending to radiate across the costal margin and the back. At nine and a half minutes (70.9 cc), the infusion was stopped. Pressure fell to 180 mm saline over the course of the next five minutes. Pain was never severe and subsided promptly after stopping the infusion. No evidence of recurrent pancreatitis was indicated by enzyme studies, and the T-tube was removed on the following day. For five years, the patient has remained well and free of digestive complaints.

Histologic examination of the gallbladder showed chronic cholecystitis. Biopsy of the papilla showed acute and chronic inflammation.

Case No. 2. AB No. 1488223

The patient, a man aged fifty-nine, had gallstones discovered by x-ray ten years previously. At that time, he had two severe attacks of epigastric pain. Subsequently, he avoided fats and was essentially well until two days before admission when he had nausea and epigastric pain with radiation to the back. On the day before entry, he had a shaking chill and developed mild icterus. On admission, he was toxic, severely ill, mildly icteric, and proved to have septicemia (anaerobic streptococci). There was generalized abdominal tenderness and rebound tenderness. All signs were maximum in the right upper quadrant.

The patient was hydrated, given antibiotics, and operated on the second day following admission. An acutely inflamed, partially intrahepatic gallbladder was encountered, with a stone impacted in the ampulla. The gallbladder contained "white" bile which cultured *E. coli*. Pancreas was normal. The stone was removed, cholecystostomy was performed, and the patient made a satisfactory recovery.

On the tenth day following operation, a tube cholangiogram demonstrated no residual stones. There was flow into the duodenum with a duodenal diverticulum demonstrated adjacent to the ampulla. On the thirteenth day, cine-cholangiograms were done with the technique similar to that described in case number 1. Resting duct pressure was 180 mm saline and with infusion at 7.64 mL per minute, pressure rose to 310 mm after 30.5 mL, and at 37 mL, pressure reached 320 mm with reflux into the pancreatic duct (fig. 2). At this time, the patient began to complain of moderate pain in the right upper quadrant with radiation to his back.

Infusion was stopped and pressure fell to 290 mm in four minutes with relief of this pain. Patient made an uneventful recovery.

Three months later, the patient was readmitted for elective cholecystectomy. At operation, the gallbladder was found to be fibrosed and without residual stones. Common bile duct and pancreas were grossly normal. An operative cholangiogram demonstrated free flow into the duodenum, reflux into the pancreatic duct, and no residual stones. Initial and final ductal pressures were 185 mm and 200 mm of saline respectively. Patient made an uneventful recovery and has been free of digestive symptoms for two years. Histologic examination of the gallbladder revealed chronic cholecystitis.

Case No. 3. NW No. 1144024

The patient, a forty-eight-year-old male, was well until four months prior to admission when he developed moderate epigastric and precordial pain. This lasted a few hours then subsided. A similar somewhat milder attack occurred two months before entry. About twelve hours prior to admission, he developed severe epigastric and precordial pain, nausea, and weakness. The symptoms persisted, and he was admitted with a diagnosis of myocardial infarction, but studies did not confirm this suspicion. Graham test showed nonfunctioning gallbladder, and he developed mild epigastric and right upper quadrant tenderness. On the eighth hospital day, operation was performed. He was found to have a tense, acutely inflamed gallbladder, thickening and nodularity of the pancreas, but no fat necrosis. A 10 mL specimen of bile was aspirated from the gallbladder for study. Grossly, the bile was very dark and thick. Operative cholangiogram showed free flow into the duodenum, no stones, and no reflux into the pancreatic duct. Ductal pressures were 440 mm of saline initially and 280 mg millimeters at the conclusion of cholangiography.

Because of the high ductal pressure, choledochostomy was performed in addition to cholecystectomy. The papilla permitted easy passage of 3 mm and 4 mm probes, but the intrapancreatic portion of the common bile duct was narrowed and would not permit a 5 mm probe to pass. A T-tube was left in the duct and the patient made an uneventful recovery. T-tube cholangiogram seven days following operation was unremarkable except for reflux into the pancreatic duct. The patient was has remained well for two years, but he has occasional mild discomfort if he overeats, and he deliberately avoids fatty foods.

Aerobic and anaerobic cultures of the bile showed no growth. Gallbladder bile contained 12 Russell units of amylase as compared to blood amylase value of 18 units. Aliquot specimens of bile from the T-tube on two separate days during recovery showed 5/15 and 365/19 bile/blood amylase ratios (Russell units) respectively.

During histological examination, the gallbladder showed acute and chronic inflammation. There were no stones in the gallbladder or the common bile duct. In addition, he had gross anatomical evidence of chronic pancreatitis.

In addition to the three cases described above, fifty-two other patients have been operated upon using the same technique as in case number 3. The series included nine males and forty-three females; ages ranged from twenty-eight to seventy-nine years. All had gallstones and chronic cholecystitis except for four who had acute cholecystitis and who are listed in table IV for a separate study.

Of this group of fifty-two cases, only eight patients demonstrated reflux of pancreatic juice into the biliary tract as evidence by bile amylase levels greater than simultaneous blood levels. Five of these also had cholangiographic evidence of pancreatic duct reflux. Four of the eight had macroscopic evidence of chronic pancreatitis. Only one of the eight had a positive culture from bile aspirated from the gallbladder.

On the other hand, of forty-four patients whose bile/blood amylase ratios were less than one (no pancreatic juice mixed in the bile, sixteen showed cholangeographic evidence of pancreatic reflux and five of these demonstrated macroscopic changes of chronic pancreatitis. Only three of the remaining twenty-eight patients in this group showed evidence of chronic pancreatitis. Only six of forty-four patients in this category demonstrated positive bile cultures.

It is of some interest that of the whole series, the patient demonstrating the highest bile/blood amylase ratio (4,320/13) also had pancreatic reflux on cholangeography but had a normal pancreas on gross examination. Histologically, his gallbladder showed chronic cholecystitis and cholesterolosis. There were two stones (each about 1cm diameter) in the gallbladder which had shown good function in response to Graham-Cole test. The bile was sterile.

Among patients who did not show reflux of pancreatic juice into the gallbladder, amylase levels in the bile averaged less than half the accompanying normal blood level. Thus, in this group of forty-four cases, the bile/blood amylase ratio averaged 4/10 Russell units.

Discussion

There is indisputable evidence that the liver synthesizes amylase, and that it secretes amylase into the blood and excretes amylase in low concentration in the bile (43, 105, 106, 107). Liver amylase may be structurally different from pancreatic amylase and may be released through cells of the reticuloendothelial system (59, 108). In normal dogs, bile amylase is present in trace amounts, but if the pancreatic duct is ligated with resultant elevation of serum amylase, increased amylase concentration occurs in bile as well as in the urine (202). The fact that amylase is excreted in low concentration in hepatic bile was well-known to Oppenheimer, Popper, and Westfall (138, 144, 185). Popper, in fact, studied bile and blood amylase values in nineteen clinical cases and deduced that if bile amylase exceeded the activity of serum amylase, it was clear evidence of reflux of pancreatic juice into the biliary system (144). The current study seems to support that conclusion.

Is pancreatic-biliary or biliary-pancreatic reflux abnormal? Given the premise that it is possible to occur in the vast majority of human subjects, it should hardly be called abnormal. In fact, from the teleological point of view, this might be considered as an adaptation to intermittent feeding since the gallbladder may, in effect, serve as a storehouse for pancreatic secretions as well as bile. This view is supported by the clinical observation that there is no close correlation between the concentration of pancreatic enzyme in the gallbladder and the severity of biliary tract disease. Flexner, in considering the mixing of bile and pancreatic juice, concluded that the colloid suspension system of bile tended to protect against activation of the pancreatic enzymes (53). Experimental hookups to divert all the bile through the pancreatic ductal system or all the pancreatic secretions through the bile duct and gallbladder have produced remarkably little inflammatory response; although, if they are continued over long periods, there tend to be progressively severe inflammatory and fibrotic changes (128, 158, 195). On the other hand, experimental designs mixing bile and pancreatic juice and adding the factor of partial obstruction regularly produce inflammation and fibrotic changes in both the pancreas and the gallbladder and may produce stones in the latter (15, 41, 45, 67, 121, 152, 183, 196). Concentrated bile alone is cytotoxic and tends to initiate chronic cholecystitis under conditions of stasis (197). Of the various pancreatic enzymes involved, trypsin seems to be the most important, but its proteolytic activity in the gallbladder depends upon a critical level of concentration (182).

Popper believed that pancreatic enzymes might be incubated with the bile in the gallbladder with enzyme activation and chronic inflammatory changes in that organ and then be emptied through the common duct and reflux into the pancreatic duct in an activated state to set off pancreatitis (144). The possibility of such a chain of events has been corroborated experimentally by Elliott, Williams, and Ellinger (47).

One may conclude that the intermixture of bile and pancreatic juice is a frequent event that the organism is usually protected from damage secondary to enzyme activation, but that long continued mixture and incubation tend to promote inflammatory changes and fibrosis. If obstruction (increased time for concentration, incubation) or infection occur, there may be abnormal enzyme activation which will result in severe inflammation and tissue necrosis (29, 174). Generally, infection plays a secondary role in cholecystitis (166, 190). The theory that lymphatic spread is possible, and that infection plays a significant role in cholecystitis or pancreatitis has been experimentally refuted by Kaufman (91).

Foley and Ellison infused activated pancreatic enzyme into the hepatic ducts of dogs and observed a marked rise in alkaline phosphatase and transaminase and increase of bilirubin to two times its baseline values. Histological study of the liver showed marked passive congestion, bile stasis, and focal necrosis of liver cells (54). This is certainly reminiscent of the acute cholecystitis of childhood. Among children, the incidence of acute cholecystitis is higher in males than in females, stones occur in only about two-thirds of the cases—jaundice in approximately one-half of these cases, and common duct stones are found in less than 10 % of the cases. The disease tends to be acute, fulminant, and associated with signs of systemic toxicity. Chronic cholecystitis is uncommon among children (75, 119, 146, 180, 193).

Acalculous cholecystitis in adults and acute cholecystitis following unrelated operations or injuries are also unusual in sex distribution—males having this disease somewhat more frequently than females. Many of these attacks occur in patients who have had no previous digestive symptoms (25, 71, 115, 126, 157, 165, 177). Arianoff found 70% of such cases to have pancreatic enzymes in the gallbladder bile (7). Cases with bile-stained peritoneal fluid would seem to be examples of Popper's bile peritonitis without perforation secondary to increased permeability of the biliary tract following enzymatic injury (100, 144).

Also, in the population at large, females predominate in demonstrating gallstones and biliary tract disease (99). The importance of pregnancy as an etiologic factor is emphasized by the high incidence of gallstones among young women who have been pregnant (96, 171).

Acute cholecystitis is a more common manifestation of the disease in this age group than among older patients (5). In part, gallstone formation in pregnancy may be due to increased cholesterol elevation in the bile (60). Thus, cholesterol stones may be induced in rabbits by administration of estrogen and progesterone (77). Also, female mice excreted cholesterol in the bile at a higher rate than males (140). The evidence as to whether the composition of human bile is altered during pregnancy is conflicting (94,154).

A second major factor in gallstone formation in pregnancy may be stasis in the gallbladder. Mann and Higgins showed diminished emptying of gallbladders of pregnant animals in response to a fatty meal as compared to controls (118). Among 390 cesarean sections at term, Potter found 75% of gallbladders large, atonic, globular, and distended with thick, black viscous bile. Only 2% of 390 cases had positive bile cultures (148). Thureborn has commented upon stasis as an etiologic agent in gallstone formation. (178) Although there do not seem to be any direct observations bearing on the point, the presence of pancreatic enzymes in the gallbladder should contribute to the formation of stones during pregnancy-induced stasis. Acute pancreatitis during pregnancy occurs infrequently (26, 93).

The increased incidence of gallstone formation after vagotomy is now established and this occurs with about equal frequency in both males and females (27). Both decreased muscular tone in the gallbladder and increased bile pressure may contribute to bile stasis in the postvagotomy state (61, 160, 192). The synergistic action of pancreatic enzymes incubating in the atonic gallbladder may contribute to chronic inflammation and stone formation.

Summary

Review of the pertinent literature and analysis of fifty-five clinical cases with study of the dynamics of the biliary-pancreatic ductal system has provided firm evidence that biliary-pancreatic or pancreatic-biliary reflux is a physiologic and probably frequent occurrence in humans. Ordinarily, this does not result in activation of proteolytic or lipophilic enzymes, and there are no resulting physical changes or symptoms. Over a prolonged period, minor episodes of enzyme activation may result in mild inflammatory changes and slight fibrosis. However, if relatively high concentrations of pancreatic juice and bile are incubated in the gallbladder under conditions of stasis and/or obstruction, activation of the enzymes may occur with subsequent enzymatic digestion of the tissues of the biliary-pancreatic tract, and, in severe cases, of the surrounding tissues. All degrees of this reaction may occur, varying from slight inflammation to frank necrosis and gangrene, and it may involve any or all areas of the entire biliary-pancreatic drainage system. Bacterial infections may occur in association with these episodes of inflammation, but they are usually secondary. However, under certain conditions, bacterial infection may act synergistically to activate the pancreatic enzymes.

Symptoms of pain, nausea and vomiting, and signs of fever, leukocytosis, dehydration, and systemic toxicity may occur with these episodes even in the absence of gallstones, and they are indistinguishable from similar symptoms which occur among patients with biliary colic and acute cholecystitis or acute pancreatitis in the presence of stones.

Following cholecystectomy, there is normally no significant dilatation of the common bile duct in humans. If obstruction distal to the confluence of the biliary and pancreatic ducts should result in significant dilatation of common bile duct with formation of a bile lake, similar enzymatic-digestive-inflammatory changes might be expected to occur. Symptoms of this reaction could hardly be distinguishable from those of the same process with the gallbladder in situ.

Although there is no evidence that this mixture of pancreatic juice with bile in the gallbladder per se results in the formation of stones, changes in the colloid properties of bile and in the mucosa of the gallbladder secondary to activation of proteolytic enzymes create conditions known to favor cholesterol precipitation. During pregnancy, the higher rate of cholesterol excretion in the bile and relative stasis of the gallbladder would be expected to promote gallstone formation if minor episodes of proteolytic enzyme activation were to occur within the gallbladder.

Although there are relatively few observations noted in the literature, it is established that the liver excretes amylase in low concentration in the bile. The level of amylase in both hepatic and gallbladder bile is lower than the concomitant amylase level in blood serum, and the finding of a bile/blood amylase ratio greater than one is a reliable index of reflux of pancreatic juice into the biliary tree.

There is a remarkably complex interrelationship in control, function, and feedback among the organs of the upper digestive tract, particularly stomach, duodenum, liver, gallbladder, and pancreas. These mechanisms are incompletely understood, but there is sufficient information to indicate that functional or organic variation from normal in one of these organs tends to be reflected in changes among the others. Specifically, the mechanisms which will promote peptic ulceration of the stomach and duodenum also operate to create disturbed function of the pancreas and gallbladder, including psychological and emotional stress.

It may be questioned whether the major contribution of cholecystectomy is the removal of the contained gallstones. The evidence suggests that the more important effect is to remove the site of bile-pancreatic juice incubation and the muscular motor organ which propels this incubated mixture into the hepatic radicals and pancreatic duct under conditions of organic or functional terminal common bile duct obstruction.

There is indisputable evidence that the liver synthesizes amylase, and that it secretes amylase into the blood and excretes amylase in low concentration in the bile. (43, 105, 106, 107) Liver amylase may be structurally different from pancreatic amylase and may be released through cells of the reticular-endothelial system. (69, 108) In normal dogs, bile amylase is present in trace amounts, but if the pancreatic duct is ligated with resultant elevation of serum amylase, increased amylase concentration occurs in bile as well as in the urine. (202) The fact that amylase is excreted in low concentration in hepatic bile was well known by Oppenheimer, Popper, and Westfall. (138, 144, 185) Popper, in fact, studied bile and blood amylase values in 219 clinical cases and deduced that if bile amylase exceeded the activity of serum amylase, it was clear evidence of reflex reflux of pancreatic juice into the biliary system. (144) This current study seems to support that conclusion.

Is pancreatic-biliary or biliary-pancreatic reflux abnormal? Given the premise that it is possible to occur in the vast majority of human subjects, it should hardly be called abnormal. In fact, from a teleological point of view, this might be considered as an adaptation to intermittent feeding since the gallbladder may, in effect, serve as a storehouse for pancreatic secretions as well as bile. This view is supported by the clinical observation that there is no close correlation between the concentration of pancreatic enzymes in the gallbladder and the severity of biliary tract disease. Flexner in considering the mixing of bile and pancreatic juice, concluded that the colloidal suspension system (bile) tended to protect against activation of the pancreatic enzymes. (53.) Experimental hookups to divert all the bile through the pancreatic ductal system for all the pancreatic secretions to the bile ducts and gallbladder have produced remarkably little inflammatory response, although if this is continued over long periods there tend to be progressively severe inflammatory and fibrotic changes. (128, 156, 195) On the other hand, experimental designs mixing bile and pancreatic juice and adding the factor of partial obstruction regularly produces inflammatory and fibrotic changes in both the pancreas and gallbladder, and may produce stones in the latter. (15, 41, 45, 67, 121, 152, 183, 196) Concentrated bile alone is cytotoxic and tends to initiate chronic cholecystitis under conditions of stasis. (197) Of the various pancreatic enzymes involved, trypsin seems to be the most important, but its proteolytic activity in the gallbladder depends on a critical level of concentration. (182)

Popper believed that pancreatic enzymes might be incubated with the bile in the gallbladder with enzyme activation and chronic inflammatory changes in that organ, and then be emptied through the common duct and reflux into the pancreatic duct in an activated state to set off pancreatitis. (144). The possibility of such a chain of events has been corroborated experimentally by Elliott, Williams and Salinger. (47)

In summary, one may conclude that the inter-mixture of bile and pancreatic juice is a frequent event, that the organism is usually protected from damage secondary enzyme activation, but that long-continued mixture and incubation tends to promote inflammatory changes and fibrosis. If obstruction (increased time for concentration, incubation) or infection occur, there may be abnormal enzyme activation with resultant severe inflammation and tissue necrosis. (29, 174) Generally, infection plays a secondary role in cholecystitis. (166, 190). The theory that lymphatic spread of infection plays a significant role in cholecystitis or pancreatitis has been experimentally refuted by Kaufman. (91).

Foley and Ellison infused activated pancreatic enzyme in the hepatic ducts of dogs and observed a marked rise in alkaline phosphatase and transaminase and increase of bilirubin to two times his base line values. Histological study of the liver showed marked passive congestion, bile stasis and focal necrosis of liver cells. (54) this is certainly reminiscent of the acute cholecystitis of childhood.

Among children, the incidence of acute cholecystitis is higher in males than in females, stones occur in only about two thirds of the cases, jaundice in approximately one half of the cases, and common duct stones are found in less than 1/10 of the cases. The disease tends to be acute, fulminant, and associated with signs of systemic toxicity. Chronic cholecystitis is uncommon among children. (75, 119, 146, 180, 193)

Acalculus cholecystitis in adults and acute cholecystitis following unrelated operations or injuries are also unusual in sex distribution with males having this disease somewhat more frequently than females. Many of these attacks occur in patients who have had no previous digestive symptoms. (25, 71, 115, 126, 157, 165, 177) Arianoff found 70% of such cases to have pancreatitis enzymes in the gallbladder bile. (7) Cases with bile-stained peritoneal fluid would seem to be examples of Pepper's "bile peritonitis without perforation" secondary to increased permeability of the biliary tract following enzymatic injury. "(100, 144)

In the population at large, there is a clear predominance of females over males in the incidence of gallstones and biliary tract disease. (99) The importance of pregnancy as an etiologic factor is emphasized by the high incidence of gallstones among young women who have been pregnant. (96, 171)

Acute cholecystitis is a more common manifestation of the disease in this age group than among older patients. (55) In part, gallstone formation in pregnancy may be due to increased cholesterol excretion in the bile. (60) Thus, cholesterol stones may be induced in rabbits by administration of estrogen and progesterone. (77) Also, female mice excrete cholesterol in the bile at a higher rate than males. (140) Thus, the evidence as to whether the composition of bile is altered during pregnancy is conflicting. (94, 191 54)

A second major factor in gallstone formation in pregnancy may be stasis in the gallbladder. Mann and Higgins showed diminished emptying of gallbladders of pregnant animals in response to a fatty meal as compared to controls. (118) Among 390 cesarean sections at term Potter found 75% of gallbladders large, atonic, globular and distended with thick, black viscous bile. Only 2% of 300 cases had positive bile cultures. (148) Thureborn has commented upon stasis as an etiologic agent in gallstone formation. (178) Although there do not seem to be any direct observations bearing on the point, the presence of pancreatic enzymes in the gallbladder should contribute to the formation of stones during pregnancy-induced stasis. Acute pancreatitis during pregnancy occurs infrequently. (26, 93)

The increased incidence of gallstone formation after vagotomy is now established and this occurs with about equal frequency in both males and females. (27) Both decreased muscular tone in the gallbladder and increased bile pressure may contribute to bile stasis in the post-vagotomy state. (61, 160, 192) The synergistic action of pancreatic enzymes incubating in the atonic gallbladder may contribute to chronic inflammation and stone formation.

Conclusions

Review of the pertinent literature and analysis of 55 clinical cases with study of the dynamics of the biliary-pancreatic ductal system has provided firm evidence that biliary pancreatic or pancreatic-biliary reflux is a physiologic and probably frequent occurrence in humans. Ordinarily, this does not result in activation of proteolytic or lipolytic enzymes and there are no resulting physical changes or symptoms. Over a prolonged period, minor episodes of enzyme activation may result in mild inflammatory changes and slight fibrosis. However, if relatively high concentration of pancreatic juice and bile are incubated in the gallbladder under conditions of stasis or obstruction, activation of the enzyme may occur with subsequent enzymatic digestion of the tissues of the biliary-pancreatic tract and, in severe cases, of the surrounding tissues. All degrees of this reaction may occur varying from slight inflammation to frank necrosis and gangrene, and may involve any or all areas of the entire biliary-pancreatic drainage system. Bacterial infections may occur in association with these episodes of inflammation-but they are usually secondary. However, under certain conditions bacterial infection may act synergistically to activate the pancreatic enzymes.

Symptoms of pain, nausea and vomiting, and signs of fever, leukocytosis, dehydration and systemic toxicity may occur with these episodes even in the absence of gallstones, and they may be indistinguishable from similar symptoms which occur among patients with "biliary colic" and acute cholecystitis or acute pancreatitis in the presence of stones.

Following cholecystectomy there is normally no significant dilatation of the common bile duct in humans. If obstruction distal to the confluence of the biliary and pancreatic duct should result in significant dilatation of the com

mon bile ducts with formation of a "bile lake", similar enzymatic-digestive-inflammatory changes might be expected to occur. Symptoms of this reaction could hardly be distinguishable from those of the same process with the gallbladder present in situ.

Although there is no evidence that the mixture of pancreatic juice with bile in the gallbladder per se results in the formation of stones, changes in the colloid properties of bile and the mucosa of the gallbladder secondary to activation of proteolytic enzymes create conditions known to favor cholesterol precipitation. During pregnancy, higher rates of cholesterol excretion in the bile and relative stasis of gallbladder bile would be expected to promote gallstone formation if minor episodes of proteolytic enzyme activation were to occur within the gallbladder.

Although there are relatively few observations noted in the literature, it is established that the liver excretes amylase in low concentration in the bile. The level of amylase in both hepatic and gallbladder bile is lower than the amylase level in blood serum; and the finding of bile/blood amylase ratio greater than 1 is a relatively reliable index of reflux of pancreatic juice into the biliary tree.

There is a remarkably complex relationship in control, function and feedback among the organs of the upper digestive tract; particularly, the stomach, duodenum, liver, gallbladder and pancreas. These mechanisms are incompletely understood, but there is sufficient information to indicate that functional or organic variations from normal in one of these organs tends to be reflected in changes among the others. Significantly, the mechanisms which promote ulceration of the stomach and duodenum also operate to create disturbed function of the pancreas and the gallbladder-including psychological and emotional stress. Significantly, it may be questioned whether the major contribution of cholecystectomy is the removal of the contained gallstones. The evidence suggests that the more important effect is to remove the site of bile-pancreatic juice incubation and, at the same stroke, remove the muscular motor organ which propels this incubated toxic mixture into the hepatic radicals and pancreatic duct under conditions of organic or functional or organic terminal common bile duct obstruction.

It may be asserted with certainty that the common bile duct has smooth muscle fibers in its wall, and that it exhibits peristaltic contractions of crescendo intensity and frequency with increasing intraductal pressure. Beginning at approximately 650 mm saline pressure, these contractions are accompanied by sharp pain radiating toward the mesial edge of the left scapula. This pain gradually subsides as common duct pressure and peristalsis subside.

Finally, gallstone formation is always intermittent—never continuous. Multiple gallstones, if present, always occur in "families" of like-sized stones. There is never a smooth, gradual sizing progression of individual gallstones.

Table I
Factors Controlling Tonus of Sphincter of Oddi
Contraction Relaxation

Pharmological

	Contraction	Relaxation
Local	0.1 HCL (6,37,38); Ethanol (33.42)	25% solution magnesium sulfate (35)
Systemic	Morphine (37, 38, 142); Ethanol (142)	Nitroglycerin, amyl nitrite (37, 142)
	Parasympathomimetic drugs and ergotamine (172)	Sympathomimetic drugs, atropine, penthonium, succinyl choline (172)

Neurological

	Contraction	Relaxation
Local	Sudden distention common bile duct (6, 97, 169)	Reflex response to pressure wave in antrum (20, 199)
	Gastroduodenal disconnection (200)	Electrical and mechanical stimulation of stomach, duodenum, or jejunum (199)
Central	Anxiety, emotional tension (37, 38)	Reassurance, calm (37)
	Vagal stimulation (172)	Section hepatic plexus anterior vagus nerve (163)
	Fasting (48)	Sight of food (48)
Hormonal	Secretin (Colletti, 87)	Cholecystokinin (20, 87, 179)

Table II
Factors Regulating Secretion of Bile and Pancreatic Juice

	Bile	Pancreatic Juice
Local	pH-sensitive receptor cell sends impulse to sell secreting secretin (170)	pH-sensitive receptor cell sends impulse to sends secreting secretive and pancreozymin (170)
Central		
Vagus		Increases enzyme content of secretin stimulated juice (111) Increases volume of enzyme-rich juice (65, 161)
Sympathetic		question slight decrease in volume and enzyme content (161)
Hormonal		
Gastrin	Increased volume and bilirubin content (28, 84)	Increased volume and enzyme content (51, 149, 198)
Secretin	Increased volume and bilirubin content (1, 2, 10)	Marked rapid increase in volume, bicarbonate content (1, 10, 57)
Cholecystokinin-pancreozymin (86)		Increased volume, enzyme-rich juice (9, 63)
Insulin		Increased volume, enzyme-rich juice (56)
Glucagon		Profound decrease in volume and enzyme content (201)
Mechanical		Gastric distention increases volume and enzyme content (question vagal reflex) (110, 189)

Pharmacological: anticholinergic drugs decrease volume and bile salt content (36); anticholinergic drugs decrease volume and enzyme content of pancreatic juice (36)

Aspirin Decholin: bile salts, zanchol increase volume, and bile salt content (13, 36, 150)

Both secretin and cholecystokinin inhibit gastric acid secretion (40, 58, 83, 85, 90, 112, 129, 130)

Table III
Enzyme Studies—MR No. 690860

Time		Blood		Bile	
		*Amylase	+Lipase	*Amylase	+Lipase
2/14/65	2:00 PM	--	--	12	0
2/15/65	9:00 AM	28	1.72	12	0.5
	11:00 AM		Cinefluoro-cholangiography		
	4:00 PM	60	3.22	380	37.8
2/16/65	12:00 AM	33	1.78	8600	80.4 (Aliquot 4:00–12:00 PM)
5/11/65	9:00 AM	12	0.70	--	--
	10:00 AM		Cinefluoro-cholangiography		
	11:00 AM	9	0.40	--	--

*Russell units
+Cubic centimeters per milliliter

Table IV
Analysis of Cases with Acute Cholecystitis

Case	Bile/ Blood *	Pancreatic reflux	Bact.	Pancreas Histology	Gallbladder Histology
MR No. 0690860	>1	+	Sterile	Acute Inflam. w/fat necrosis	Ac/Chron. cholecystitis, stones—not impacted
AR No. 1488223	>1	+	E. Coli., Alpha hem. Strep	Normal	Ac./Chron. Cholecystitis, white bile, impacted stone
NW No. 1141024	>1	+	Sterile	Chronic Inflam.	Ac./Chr. Cholecystitis—no stones
ES No. 0348610	>1	+	Sterile, Chronic Inflam.	Ac. Gangrenous/ chron.	Cholecystitis. Stones, not impacted
MP No. 1270559	>1	+	Staph. Aureus	Normal	Ac. Gangrenous/ chron. cholecystis, 1 stone forming cast of the entire gallbladder
TF No.0785807	<1		Sterile	Chronic inflam.	Ac./Chron. cholecystitis, stones—not impacted
RB No. 1372624			Staph. Aureus	Normal	Subacute gangrenous cholecystitis, stone impacted in ampulla

*Bile-blood amylase ratio >1 indicates reflux of pancreatic juice into the biliary tree

References

1. Agren, G. and Lagerlof, H. "The Pancreatic Secretion in Man after Intravenous Administration of Secretin." *Acta Med. Scand.* 90:1–29, 1936.

2. Agren, G. and Lagerlof, H. "The Biliary Response in the Secretin Test." *Acta Med. Scand.* 92:359–366, 1937.

3. Anderson, M. C., Mehn, W. H., and Method, H. L. "Evaluation of Common Channel as Factor in Pancreatic or Biliary Disease." *Ann. Surg.* 151:37–390, 1960.

4. Anderton, J. L., Finlayson, N. D. C., Nimmo, J., Smith, A. F., and Shearman, D. J. C. "Acid Base Change Due to Secretin and Administration in Normal Patients and in Patients with Pancreatic Disease." *Scand. J. Gastroent.* 4:335–339, 1969.

5. Archibald, E. "Ideas Concerning the Causation of Some Cases of Pancreatitis." *Canad. J. M. Surg.* 33:263–268, 1913.

6. Archibald, E. and Gibbons, E. C. "Further Data Concerning the Experimental Production of Pancreatitis." *Ann. Surg.* 74:426–433, 1921.

7. Arianoff, A. A. "Les Cholecystitis Alithiasiques Vues par le Chirugien." *Acta Gastroenterol. belg.* 25:432–452, 1962.

8. Aronson, H. G. and Andrews, B. "Nonbacterial Cholecystitis: The Mechanism of Acidification of Bile in the Gallbladder." *Proc. Soc. Exp. ER. Biol. and Med.* 33:89–91, 1935.

9. Banwell, J. G., Northam, B. E., and Cooke, W. T. "Secretory Response of the Human Pancreas Continuous Intravenous Infusion of Pancreozymin-Cholecystokinin." *Gut.* 8:380–387, 1967.

10. Baylis, W. M. and Starling, E. H. "The Mechanism of Pancreatic Secretion." *Jour. of Phys.* 28:325–353, 1902.

11. Beneventano, T. C., Jacobson, H. G., Hurwitt, E. S., and Schein, C. J. "Cinecholangiomanometry: Physiologic Observations." *Amer. J. Roentgen.* 100:673–679, 1967.

12. Bergh, G. S. "Sphincter Mechanism Common Bile Duct in Human Subjects." *Surgery.* 11:299–330, 1942.

13. Best, R. R. "The Quantitative and Qualitative Control of Bile Flow and Its Relation to the Biliary Tract Surgery." *Ann. Surg.* 128:348–361, 1948.

14. Best, R. R. and Hicken, N. P. "Biliary Dyssynergia: Physiological Function of the Common Bile Duct." *Surg. Gynec. and Obstet.* 61:721–724, 1935.

15. Bisgard, J. The. and Baker, C. P. "Studies Relating to the Pathogenesis of Cholecystitis, Cholelithiasis, and Acute Pancreatitis." *Ann. Surg.* 112:1006–1034, 1940.

16. Blad, A. "Studies uber Gallenperitonitis ohne Perforation der Gallenwege." *Arch. Klin. Chir.* 109:101–120, 1917–1918.

17. Bouchier, I. A. D., Cooperband, S. R., and El Yodsi, B. Then. "Call on Mucous Substances and Viscosity of Normal and Pathological Human Bile." *Gastroenterology.* 49:343–353, 1965.

18. Boyden, E. A. "The Sphincter of Oddi in Man and Certain Representative Mammals." *Surg.* 1:25–3, 1937.

19. Boyden, E. A. "The Anatomy of the Choledocho-Duodenal Junction in Man." *Surgery, Gynecology and Obstetrics.* 104:641–652, 1957.

20. Burton, P., Harper, A. A., Howatt H. T., Scott, J. E., and Varley, H. "The Use of Cholecystokinin to Test Gallbladder Function in Man." *Gut.* 1:193–204, 1960.

21. Butler, T. J. "A Study of the Pancreatic Response to Food after Gastrectomy in Man." *Gut.* 1:55–61, 1960.

22. Byrne, J. J. and Boyd, T. F. "Hyperamylasemia in Intestinal Obstruction and Its Relationship to Pancreatitis." *American Journal of Surgery.* 105:720-729, 1963.

23. Cameron, A. L. and Noble, J. F. "Reflux of Bile of the Duct of Wirsung Caused by an Impacted Biliary Calculus. *J. A. M. A.* 82:1410–1414, 1924.

24. Caroli, J. P., Porcher, P., Pequignot, G., and Delattre, M. "Contribution of Cinoradiography to Study the Function of the Human Biliary Tract." *American Journal of Digestive Disease.* 5:677–696, 1960.

25. Carter, R. F., Greene, C. H., Twiss, J. R., and Hotz, R. "Etiology of Gallstones." *Archives of Surgery.* 39:691–710, 1939.

26. Cassell, W. J. Jr. and Malewitz, E. C. "Acute Pancreatitis in Pregnancy." *J. A. M. A.* 142:1139–1140, 1950.

27. Clave, R. A. and Gaspar, M. R. "Occurrence of Gallbladder Disease after Vagotomy." *American Journal of Surgery.* 118:169–174, 1969.

28. Cole, G. J. and Connell, G. M. "Gastrin, Acid, and Bile." *Gut.* 9:638–640, 1968.

29. Colp, R. and Doubilet, H. "Sphincter Oddi: Clinical Significance of Pancreatic Reflux." *Annals of Surgery.* 108:243–262, 1938.

30. Colp, R., Gerber, I. E., and Doubilet, H. "Acute Cholecystitis Associated with Pancreatic Reflux." *Annals of Surgery.* 103:67–76, 1936.

31. Colp, R., Gerber, I. E., and Doubilet, H. "Acute Cholecystitis Associated with Pancreatic Reflux." *Annals of Surgery.* 103:67–76, 1936.

32. Daniels, B. T., McGlove, F. B., Joo, H., and Sawyer, R. B. "Changing Concepts of Common Bile Duct Anatomy and Physiology." *J. A. M. A.* 178:394–397, 1961.

33. Davis, A. E. and Payroll the, Bar. C. "The Effects of Ethyl Alcohol on Pancreatic Exocrine Function." *Med. J. Australia.* 2:757–760, 1966.

34. Demand, H. A. and Kuckuck, P. "Correlation Analysis of Duodenal Aspirates after Secret in Administration." *Scandinavian Journal Gastroenterology.* 4:329–334, 1969.

35. Doubilet, H. and Colp, R. "Resistance of the Sphincter of Oddi in the Human." *Surgery, Gynecology and Obstetrics.* 64:622–633, 1937.

36. Doubilet, H. and Fishman, L. "Human Biliary-Pancreatic Secretion." *American Journal of Gastroenterology.* 35:499–512, 1961.

37. Doubilet, H. and Mulholland, J. H. "The Surgical Treatment of Recurrent Acute Pancreatitis by Endocholedochal Sphincterotomy." *Surgery, Gynecology and Obstetrics.* 86:295–308, 1948.

38. Doubilet, H. and Mulholland, J. H. "Recurrent Acute Pancreatitis: Observations on Etiology and Surgical Treatment." *Annals of Surgery.* 128:609–636, 1954.

39. Doubilet, H., Poppel, M. H., and Mulholland, J. H. "Pancreatography." *Radiology.* 64:325–329, 1955.

40. Dragstedt, L. R. "Duodenal Inhibition of Gastric Secretion." *Amer. J. Surg.* 117:841–848, 1969.

41. Dragstedt, L. R., Hammond, H. E., and Ellis, J. C. "Pathogenesis of Acute Pancreatitis." *Arch, Surg.* 28:232–291, 1934.

42. Dreiling, D. A., Richman, A., and Fradkin, N. F. "The Role of Alcohol in the Etiology of Pancreatitis: A Study of the Effect of Intravenous Ethyl Alcohol on the External Secretion of the Pancreas." *Gastroenterology.* 20:636–646, 1952.

43. Dreiling, D. A., Rosenthal, W. S., Cass, M., and Janowitz, H. D. "Relationship between Blood Amylase and Factors Affecting Carbohydrate Metabolism." *American Journal of Digestive Diseases.* 4:731–736, 1959.

44. Dubois, F. S. and Kistler, G. H. "Concerning the Mechanism of Contraction of the Gallbladder in the Guinea Pig." *Proceedings Society Experimental Biology and Medicine.* :1178–1180, 1933.

45. Dunphy, J. E. and Robinson, T. M. "Effects of Incomplete Obstruction of the Common Bile Duct." *Archives of Surgery.* 83:18–26, 1961.

46. Egdahl, R. N. "Mechanism of Blood Enzyme Changes Following the Production of Experimental Pancreatitis." *Annals of Surgery.* 148:389–400, 1958.

47. Elliott, D. W., Williams, R. D., and Zollinger, R. M. "Alterations in the Pancreatic Resistance to Bile in the Pathogenesis of Acute Pancreatitis." *Annals of Surgery.* 146:669–681, 1957.

48. Elliott, D. W, and McMaster, P. D. "The Physiological Variations in Resistance to Bile Flow to the Intestine." *Journal of Experimental Medicine.* 44:151–198, 1926.

49. Elmslie, R. G., Thorpe, M. E. C., Coleman, J. V. L., Boughton, C. R., Pritchard, G. R., and Hoy, R. J. "Clinical Significance of White Bile in the Biliary Tree." *Gut.* 10:530–533, 1969.

50. Elmslie, R. G., White, T. T., and Magee, D. F. "Observation on Pancreatic Function in Eight Patients with Controlled Pancreatic Fistulas." *Annals of Surgery.* 160:937–949, 1964.

51. Emas, S., Billings, A., and Grossman, M. I. "Effects of Gastrin and Pentagastrin on Gastric and Pancreatic Secretion in Dogs." *Scandinavian Journal Gastroenterology.* 3:234–240, 1968.

52. Emerson, W. See. and Whitaker, L. R. "The Effect of Eliminating the Sphincter of the Common Bile Duct upon Emptying of the Gallbladder." *American Journal of Physiology.* 83:484–487, 1928.

53. Flexner, S. "The Constituent of the Bile Causing Pancreatitis and the Effects of Colloids upon Its Action." *Journal of Experimental Medicine.* 8:167–177, 1906.

54. Foley, J. J. and Ellison, E. H. "Infusion of Pancreatic Enzymes into Biliary Radicals of Liver." *Archives of Surgery.* 88:589–595, 1964.

55. Fosburg, R. G. "Gallstones in Young Adults: An Analysis of 178 Patients under 30 Years of Age." *American Journal of Surgery.* 106:82–88, 1963.

56. Frisk, A. R. and Walin, G. "The External Pancreatic Secretion and the Discharge of Bile during Hypoglycemia Following Intravenous Administration of Insulin. *Acta Med. Scand.* 91:170–182, 1937.

57. Gerber, M. L., Hertzog, F. J., Lee, Y. P., Field J. B., and Drapanas, T. "Effect of Secretin on Visceral Hemodynamics and Pancreatic Blood Flow." *Surg. Forum.* 17:385–387, 1967.

58. Gillespie, I. E. and Grossman, M. I. "Inhibitory Effect of Secretin and Cholecystokinin on Heidenhain Pouch Responses to Gastrin Extract and Histamine." *Gut.* 5:342–345, 1964.

59. Gilsdorf, R. B., Urdaneta, L. F., Delaney, J. P., and Leonard, A. S. "Central Nervous System Influence on Pancreatic Secretions, Sphincteric Mechanisms, and Blood Flow and Their Role in the Course of Pancreatitis." *Surgery.* 62:581–588, 1967.

60. Glenn, F. and McSherry, C. K. "Pregnancy, Cholesterol Metabolism and Gallstones." *Annals of Surgery.* 169:712–723, 1969.

61. Griffith, C. A. "Significant Functions of the Hepatic and Celiac Vagi." *American Journal of Surgery.* 118:251–253, 1969.

62. Grossman, M. I., Greenguard, H., and Ivy, A. C. "The Effect of Dietary Composition on Pancreatic Enzymes." *American J. Physiology.* 138:676–682, 1943.

63. Hanscom, D. H., Jacobson, B. H., and Littman, A. "The Output of Protein after Pancreozymin." *Annals of Internal Medicine.* 66:721–726, 1967.

64. Hansen, K. "Experimental and Clinical Studies in Etiologic Role of Bile Reflux in Acute Pancreatitis." *Acta Chir, Scand Supplement.* 375:5–102, 1967.

65. Hayama, T., Magee, D. F., and White, T. T. "Influence of Autonomic Nerves on the Daily Secretion of Pancreatic Juice in Dogs." *Annals of Surgery.* 158:290–294, 1963.

66. Hendrickson, W. F. "A Study of the Musculature of the Entire Extrahepatic Biliary System, Including that of the Duodenal Portion of the Common Bile Duct and of the Sphincter." *Johns Hopkins Hospital Bulletin.* 9:221–232, 1898.

67. Herman, R. E. and Davis, J. H. "The Role of Incomplete Pancreatic Duct Obstruction in the Etiology of Pancreatitis." *Surgery.* 48:318–329, 1960.

68. Herring, P. T. and Simpson. S. "The Pressure of Pancreatic Secretions and the Mode of Absorption of Pancreatic Juice after Obstruction of the Main Duct of the Pancreas." *Quarterly Journal Experimental Physiology.* 2:99–108, 1909.

69. Hyatt, N., Coverdale, G. M., and Bonorris, G. "Thorotrast Inhibition of Amylase Synthesis by the Isolated, Perfused Rat Liver." *Proceedings of Experimental Biology and Medicine.* 121:1122–1124, 1966.

70. Hiatt, N., and Warner, N. E. "Serum Amylase and Changes in Pancreatic Function and Structure after Ligation." *American Surgeon.* 35:30–35, 1969.

71. Hilton H. D. and Griffin, W. T. "Acute Acalculus Cholecystitis." *Surgery.* 64:1047–1048, 1968.

72. Hjorth, T. E. Contributions to the Knowledge of Pancreatic Reflux as an Etiologic Factor in Chronic Affections of the Gallbladder: Experimental Study." *Acta Chir. Scandinavia.* 96 (Supplement): 134, 1–76, 1947.

73. Holm, JP C., Edmonds, L. Age., Junior., and Baker, JP W. "Life-Threatening Complications after Operations upon the Biliary Tract." *Surgery, Gynecology and Obstetrics.* 127:241–252, 1968.

74. Hong, S. S, Magee, D. F., and Crewdson F. "The Physiologic Regulation of Gallbladder Evacuation." *Gastroenterology.* 30:625–630, 1956.

75. Hopkins, J. W., Lynn, H. B., and Dower, J. C. "Acute Non-calculus Cholecystitis in a Three-Year-Old Child." *Clin. Pediatrics.* 1:105–110, 1962.

76. Howard, F. and Yudkin, J. "Effect of Dietary Change upon the Amylase and Trypsin Activities of the Rat Pancreas." *British Journal Nutrition.* 17:281–294, 1964.

77. Imamoglu, K., Wangensteen, S. L., Root, H. D., Salmon, P. A., Griffen, W. O. Jr., and Wangensteen. O. H. "Production of Gallstones by Prolonged Administration of Progesterone and Estradiol in Rabbits." *Surgical forum.* 10:246–249, 1959.

78. Ivy, A. C. "The Physiology of the Gallbladder." *Physiology Review*. 14:1–102, 1934.

79. Ivy, A. C. and Gibbs, G. E. "Pancreatitis: A Review." *Surgery*. 31:614–642, 1952.

80. Ivy Century, A. C. and Oldberg, E. "A Hormone Mechanism for Gallbladder Contraction and Evacuation." *American Journal Physiology*. 86:599–613, 1928.

81. Jacobson, B. "Determination of Pressure in the Common Bile Duct at and after Operation." *Acta Chir. Scand*. 113:483–488, 1957.

82. Jacobson, B., Lanner, L. O., and Radberg, C. "The Dynamics of the Choledocho=Duodenal Junction Studied by Cineroentgenography during Pressure Recording." *Acta Chir. Scand*. 113:488–490, 1957.

83. Johnson L. P., Brown, J. C., and Magee, D. F. "Effect of Secretin and Cholecystokinin-Pancreozymin Extracts on Gastric Motility in Man." *Gut*. 7:52–57, 1966.

84. Jones, R. S., Powell, K. C., and Brooks, F. P. "The Role of the Gastric Antrum in the Control of Bile Flow." *Surgical Forum*. 16:386–387, 1965.

85. Jordan, B. H. Jr. and De la Rosa. "Inhibition of Gastric Secretion by Duodenal Mucosal Extracts." *Annals of Surgery*. 160:978–985, 1965.

86. Jorpes, E. and Mutt V. "Cholecystokinin and Pancreozymin, One Single Hormone?" *Acta Physiology Scandinavia*. 66:196–202, 1966.

87. Jorpes, E. and Mutt, V. "Clinical Aspects of the Gastrointestinal Hormones, Secretin, and Cholecystokinin." *Scandinavian Journal of Gastroenterology*. 4:49–57, 1969.

88. Judd, E. S. "The Recurrence of Symptoms Following Operation of the Biliary Tract." *Annals of Surgery*. 67:473–488, 1931.

89. Judd, E. S. and Mann F. "The Effect of Removal of the Gallbladder." *Surgery, Gynecology and Obstetrics*. 24:437–442, 1917.

90. Kamionkowski, M., Grossman, S., and Fleshler, B. "The Inhibitory Effect of Secretin on Broth-Stimulated Gastric Secretion in Human Subjects." *Gut*. 5:237–240, 1961.

91. Kaufman, M. "Experimental Study of the Lymphatic Theory of Pancreatitis." *Surgery Gynecology and Obstetrics*. 44:15–22, 1927.

92. Kock, N. G., a, B., and Jacobson, B. "The Influence of the Motor Activity of the Duodenum on the Pressure in the Common Bile Duct." *Panels of Surgery*. 160:950–957, 1964.

93. Langmade, C. F. and Edmondson, H. A. "Acute Pancreatitis during Pregnancy and the Postpartum Period." *Surgery, Gynecology and Obstetrics*. 92:43–52, 1951.

94. Large, A. M., Johnston, C. G., Katsuki, T., and Fachnie, H. L. "Gallstones and Pregnancy: The Composition of Gallbladder Bile in Pregnant Women at Term." *American Journal Medical Science*. 239:713–720, 1960.

95. Larry, B. C. "Studies in 204 Routine Operative: Cholangiograms." *American Journal Gastroenterology*. 46:130–134, 1966.

96. Laws, Age. L. "Gallbladder Disease in Young People." *American Surgeon*. 35:480–481, 1969.

97. Layne, J. A. and Berg, G. S. "An Experimental Study of Pain in the Human Biliary Tract Induced by Spasm of the Center of Body." *Surgery, Gynecology and Obstetrics*. 70:18–24, 1940.

98. Lichtenstein, M. E. and Ivy, A. C. "The Function of the Valves of Heister." *Surgery*. 1:38–52, 1937.

99. Lieber, M. M. "The Incidence of Gallstones and Their Correlation with Disease." *Annals of Surgery*. 135:394–405, 1952.

100. Lindberg, E. F., Grinnan, G. L. B., and Smith, L. "Acalculus Cholecystitis in Vietnam Casualties." *Annals of Surgery*. 171:152–157, 1970.

101. Longo, M. F., Hodgson, J. R., and Ferris, D. O. "Size of the Common Duct Following Cholecystectomy." *Annals of Surgery*. 165:250–253, 1967.

102. Ludwick, J. R. and Bass, P. "Contractile Flow and Electric Activity of the Extrahepatic Biliary Tract and Duodenum." *Surgery, Gynecology and Obstetrics*. 124:536–546, 1967.

103. McCutcheon, A. D. "Reflux of Duodenal Contents in the Pathogenesis of Pancreatitis." *Gut*. 5:26–265, 1964.

104. McCutcheon, A. D. "A fresh approach to the pathogenesis of pancreatitis". Gut, 9:296-310, 1968.

105. McGeachin, R. L. "Amylase Synthesis and Transport in the Isolated, Perfused Liver." *Advances Enzymatic Regulation*. 3:137–143, 1965.

106. McGeachan, R. L., Potter, B. A., and Despopoulos. A. P. "Amylase Synthesis in the Isolated, Perfused Liver." *Archives of Biochemistry*. 90:319–320, 1960.

107. McGeachan. R.L. Potter, B.A. and Despopoulos, A.P.. "Factors Affecting Amylase Output by the Isolated, Perfused Liver." *Archives of Biochemistry*. 98:89–94, 1962.

108. McGeachan, R. L. and Tabler, Asked. L. "Effect of Physical Factors on Liver Amylase Activity." *Proceedings Society Experimental Biological Medicine*. 113:1003–1006, 1963.

109. Mack, A. J. and Todd, J. K. "A Study of the Human Gallbladder Muscle in Vitro." 9:546–549, 1968.

110. Magee, D. F., Fragola, L. A., and White, T. T. "Gastric Acid and the Gastropancreatic Distention Reflex." *Gastroenterology*. 44:811–812, 1963.

111. Magee, D.F. Fragola, L.A., and White, T.T.. "Influence of Parasympathetic Innervation on the Volume of Pancreatic Juice." *Annals of Surgery.* 161:15–20, 1965.

112. Maklouf, G. M., McManus, J. P. A., and Card, W I. "Dose-Response Curves for the Effect of Gastrin II on Acid Gastric Secretion in Man." *Gut.* 5:379–384, 1964.

113. Mallet-Guy, P, Gignoux, M., Ssterpin, P., and Yoshitomi, G. "Le devenir dela voie biliafe principale apres cholecystectomie—II, Controle du tonus du sphincter dOddi apraes cholecystectomie—II, Controle du tonus du sphincter d'Oddi apres cholecystectomie experimentale." *Lyon Chir.* 64:887–897,1968.

114. Mallet-Guy, P., Gignoux, M., Sterpin, P., and Yoshitomi, G. "Le devenir dela voie biliare principale apres cholecystectomie—III, Compaarison du caliber cholledocien pre-et post-operatoire cheg 100 operes." *Lyon Chir.* 64:898–908, 1968.

115. Mandelbaum, I. and Palmer, R. M. "Post-traumatic Acalculus Cholecystitis." *Archives of Surgery.* 97:601–604, 1968.

116. Mann, F. C. and Giordano, A. S. "The Bile Factor in Pancreatitis." *Archives of Surgery.* 6:1–30, 1923.

117. Mann, F. C. and Higgins, G. M. "Effect of Pregnancy on the Emptying of the Gallbladder." *Archives of Surgery.* 15:552–559, 1927.

118. Mann, F. See., Espinosa, J., and Hyman, L J. "Acute Acalculus Cholecystitis in Childhood." *J. Pediatric Surgery.* 3:608–611, 1968.

119. Marks, C., Espinosa, J., and Hyman, L. J. "Acute Acalculus Cholecystitis in Childhood." *Journal Pediatric Surg.* 3:6C8–611, 1968.

120. Menguy, R. "Dynamics of the Biliary Tree." *American Journal of Surgery.* 107:365–366, 1964.

121. Menguy, R., Hallenbeck, G. A., Bollman, J. L., and Grindlay, J. H. "Ductal and Vascular Factors in Etiology of Experimentally Induced Acute Pancreatitis." *Archives of Surgery.* 74:881–889, 1957.

122. Menguy, R. Hallenbeck, G.A., Bollman, J.L., and Grindlay, J. H.: "Intraductal Pressures and Sphincteric Resistance in Canine Pancreatic and Biliary Ducts after Various Stimuli." *Surgery, Gynecology and Obstetrics.* 106:306–320, 1958.

123. Meyer, W. "Chronic Cholecystitis without Stones: Diagnosis and Treatment." *Annals of Surgery.* 74:439–448, 1921.

124. Melbourne, B. "Excretory Ducts of the Pancreas in Man, with Special Reference to Their Relations to Each Other, to Common Bile Duct, and to Duodenum." *Acta Anat.* 9:1–34, 1950.

125. Mitchell, W. T. and fl, Are. B. "The Pressure of Bile Secretion during Chronic Obstruction of the Common Bile Duct." *Johns Hopkins Hospital Bulletin.* 27:78–79, 1916.

126. Munster, A. M. and Brown, J. R. "Acalculus Cholecystitis." *American Journal of Surgery.* 113:730–734, 1967.

127. Myers, R. N., Haupt, G. L., Birkhead, N. C., and Deaver, J. M. "Cindy Fluoro Graphic Observations of Common Bile Duct Physiology." *Annals of Surgery.* 156:442–449, 1962.

128. Najarian, J. S., Hine, D. E., Whitrock, R. M., and McCorcle, H. J. "Effect of Pancreatic Secretions on the Gallbladder." *Archives of Surgery.* 74:890–896, 1957.

129. Nakamura, N., Nakajima, S., and Magee, D. F. "Action of Pancreozymin on Gastric Acid Secretion." *Gut.* 9:405–410, 1968.

130. Nakayama, S. and Fukuda, H. "Effects of Cholecystokinin Preparation on the Movement of the Stomach and Small Intestine." *Japanese Journal Physiology.* 16:185–193, 1966.

131. Nebesar, R. A., Pollard, J. J., and Potsaid, M. S. "Cinecholangeography: Some Physiologic Observations." *Radiology.* 86:475–479, 1966.

132. Oddi, R. "D'une Disposition a Sphincter Speciale de L'ouverture Du Canao Choledoque." *Arch. de biol.* 8:317–322, 1887.

133. Oddi, R. "Effects de l'extirpationde l vesicule biliare." *Arch. Ital. di 'biol.* 10:425, 1888.

134. Oddi, R. "Sulla Tonicita Dello Sfincters del Coledoco." *Arch per le Sci.Med. Torino.* 12:333–339, 1888.

135. Ono, K., Watanabe, N., Suzuki, K., Tsuchida, H., Sugiyama, Y., and Abo, M. "Electrophysiologic and Cinefluorographic Observations of Bile Flow Mechanism in Man." *Surgical Forum.* 18:416–417, 1967.

136. Ono, K., Watanabe, N., Suzuki, K., Tsuchida, H., Suglyama, Yk., and Abo, N. "Bile Flow Mechanism in Man." *Arch. Surg.* 96:869–874, 1968.

137. Opie, E. "The Etiology of Acute Hemorrhagic Pancreatitis." *Johns Hopkins Hospital Bulletin.* 12:182–188, 1901.

138. Oppenheimer, C. and Kuhn R. "Lehrbuch der enzyme." *Leipzig, G. Thieme.* 325–328, 1927.

139. Parry, E. W., Hallenbeck, G. A., and Grindlay, J. H. "Pressure in the Pancreatic and Common Ducts: Values during Fasting, after Various Meals, and after Sphincterotomy; an Experimental Study." *Archives of Surgery.* 70:757–765, 1955.

140. Pedreira F. and Tepperman, G. "Bile Flow Rate and Cholesterol Content in Mice Fed a Gallstone-Inducing Diet." *American Journal Physiology.* 206:635–640, 1964.

141. Pikula, J. V. and Dunphy, J. E. "Some Effect of Stenosis of the Terminal Common Bile Duct on the Biliary Tract and Liver." *N. E. J. M.* 260:315–318, 1959.

142. Pirolla, R. C. and Davis, A. P. "Effects of Ethyl Alcohol on Sphinteric Resistance at the Choledocho-Duodenal Junction in Man. *Gut.* 9:557–560, 1968.

143. Popper, H. L. "Zur Entstehung der perforationslosenGalligen Peritonites, Zentralbl." *F. Chir.* 57:2837–2841, 1930.

144. Popper, H. L. "Pankreassaft in den Gallenwegen. Seine pathogenetische BedeutungFur die Entstehung der akuten Pancreaserkrankurgen." *Arch. f. Klin. Chir.* 175:660–695, 1933.

145. Popper, H. L. and Necheles, H. "Pathways of Enzymes into the Blood in Acute Damage of the Pancreas." *Proceedings Society Experimental Biology and Medicine.* 43:220–222, 1940.

146. Potter, A. H. "Biliary Disease in Young Subjects." *Surgery, Gynecology and Obstetrics.* 46:795–808, 1928.

147. Potter, J. C. and Mann, F. C. "Pressure Changes in the Biliary Tract." *American Journal Medical Science.* 171:202–217, 1926.

148. Potter, M. G. "Observations of the Gallbladder and Bile during Pregnancy at Term." *J. A. M. A.* 106:1070–1074, 1936.

149. Preshow, R. N. and Grossman, M. I. "Stimulation of the Pancreatic Secretion by Extracts of the Pyloric Gland Area of the Stomach." *Gastroenterology.* 48:36–44, 1965.

150. Puestow, C. B. "The Discharge of Bile into the Duodenum." *Archives Surgery.* 23:1013–1029, 1931.

151. Quist. C, F. The Influence of Cholecystectomy on the Normal Common Bile Duct." *Acta. Chir. Scand.* 113:30–34, 1957.

152. Reid, S. E. "Effect of Pancreatic Juice on the Gallbladder." *Surgery, Gynecology and Obstetrics.* 89:160–164, 1949.

153. Riegel, C. R., Ravdin, I. S., Johnston, C. G., and Morrison, P. J. "Studies of Gallbladder Functions XIII: The Composition of the Gallbladder Bile and Calculi in Gallbladder Disease." *Surgery, Gynecology and Obstetrics.* 62:933–940, 1936.

154. Riegel, C. R., Rafdin, I. S., Morrison, P. J., and Potter, M. J. "Studies of Gallbladder Function: XI. The Composition of the Gallbladder Bile in Pregnancy." *J. A. M. A.* 105:1343–1344, 1935.

155. Reinhoff, W. F. Jr. and Pickrell, K. L. "Pancreatitis: An Anatomic Study of Pancreatic and Extrahepatic Biliary Systems." *Archives of Surgery.* 51:205–219, 1945.

156. Robinson, T. M. and Dunphy, J. E. "Continuous Perfusion of Bile and Protease Activators through the Pancreas " *J. A. M. A.* 183:530–533, 1963.

157. Rosenberg, S. A. and Buchanan, J. J. "Acute Acalculus Cholecystitis Unrelated to Previous Operation." *American Surgery*. 32:319–321, 1966.

158. Roux, F. and Mc Master, P. D. "The Determination Factor in the Causation of White Stasis Bile." *Proceedings Society Experimental Biology and Medicine*. 17:159, 1920.

159. Rudick, J. and Hutchison, J. S. F. "Effects of Vagal Nerve Section on the Biliary System." *Lancet*. 1:579–581, 1964.

160. Rudick, J. and Hutchison, J.S.F. "Evaluation of Vagotomy and Biliary Function by Combined Oral Cholecystography and Intravenous Cholangiography." *Annals Surgery*. 162:234–240, 1965.

161. Sarles, H., Dani, R., Pregalin, G., Souville, C., and Figarella, G. "Cephalic Phase of Pancreatic Secretions in Man. *Gut*. 9:214–221, 1968.

162. Schein C. J., Beneventano, T. C., Rosen, R. G., Dardik. H. M., and Gliedman, M. L. "Hepatic Plexus Vagotomy as an Adjunct to Cholecystectomy." *Surgery, Gynecology and Obstetrics*. 128:241–251, 1969.

163. Schein, C. J., Beneventano, T. C., Rosen, R. G., Dardik, H. M., and Gliedman, M. L. "Hepatic Plexus Vagectomy as an Adjunct to Cholecystectomy." *Surgery, Gynecology and Obstetrics*. 128:241–251, 1969.

164. Schmidt, C. R. and IV, A. See. "The General Function of the Gallbladder: Do Species Lacking a Gallbladder Possesses Its Functional Equivalent? The Bile and Pigment Output of Various Species of Animals." *Journal Cell Physiology*. 10:365–383, 1937.

165. Schwegman, C. W. and DeMuth W. E. Jr. "Acute Cholecystitis Following Operation for Unrelated Disease." *Surgery, Gynecology and Obstetrics*. 97:167–172, 1953.

166. Scott, A. J. and Khan, G. A. "Origin of Bacteria in Bile Duct Bile." *Lancet*. 2: 790–792, 1967.

167. Shriner, W. "Studies on the Common Duct Pressures and Mechanisms." *American Journal of Gastroenterology*. 48:30–37, 1967.

168. Sinclair, I. S. R. "Observations on a Case of External Pancreatic Fistula in Man." *British Journal Surgery*. 44:250–262, 1956.

169. Sitnick, J. L., Warren, W. D., and Klien, M. R. "The Evaluation of Sphincter of Oddi Function." *Surgical Forum*. 15:373–375, 1964.

170. Slayback, J. B., Swena, E. M., Thomas, J. E. A., and Smith, L. L. "Secretion Response of the Pancreas to Topical Anesthetic Block of the Small Ball." *Surgical Form*. 17:351–353, 1966.

171. Sparkman, R. S. "Gallstones in Young Women." *Annals of Surgery*. 145:813–824, 1957.

172. Stalport, J. "A Study of the Physiopathology of the Sphincter of Oddi. (Etude par debitmetric, de la Physiopathologic oddienne)." *J. Chirug. Par*. 88:11–32, 1964.

173. Stavney, S. E. Jr. and Nagel, C. B. "Acute Cholecystitis: An Experimental Study." *Annals Surgery.* 167:689–694, 1963.

174. Stavney, L. S., Kato, T., Griffith, C. A., Nylus, L. M., and Harkins, H. N. "A Physiologic Study of Motility Changes Following Selective Vagotomy." *Journal Surgical Research.* 3:390–394, 1963.

175. Stracker, O. "Die Plica Longitudinalis Duodeni biem Menschen und bei Tieren. Sitz ungsb. d. K, Akad d.Wissench. Math-nature, K1, Wien. 3." *Abt, Hundred.* 18:375–437, 1909. 176.

176. Thistlethwaite, J. R. and Smith, D. F. Evaluation of Sphincterotomy for Treatment of Chronic Recurrent Pancreatitis." *Annals of Surgery.* 153:226–232, 1963.

177. Thompson, J. W., Ferris, D. O., and Baggenstoss, A. H. "Acute Cholecystitis Complicating Operation for Other Disease." *Annals of Surgery.* 155:489–494, 1962.

178. Thureborn, E. "On the Stratification of Human Bile and It's Importance for the Solubility of Cholesterol." *Gastroenterology.* 50:775–780, 1966.

179. Torsoli, A., Ramorino, M. L., Colagrande, C., and Demaio, G. "Experiments with Cholecystokinin." *Acta Radiol.* 55:193–206, 1961.

180. Ulin, A. W., Nosal, J. I. and Martin, W. L. "Cholecystitis in Childhood: Associated Obstructive Jaundice." *Surgery.* 31:312–326, 1952.

181. Vajoner, A., Grossling, S., and Nicoloff, D. M. "Physiologic Evaluation of Sphincter Old Plasty." *Surgery.* 62:589–595, 1967.

182. Wagner, D. E., Elliott, D. W., Endahl, G. L., and McPherson, G. T. "Specific Pancreatic Enzymes in the Etiology of Acute Cholecystitis." *Surgery.* 52:259–261, 1962.

183. Weinstein, O. H., Levin, N., and Manson, N. H. "Acute Pancreatitis (Pancreatic Necrosis). An Experimental and Clinical study, with Special Reference to the Significance of the Biliary Tract Factor." *Archives of Surgery.* 23:47–73, 1931.

184. Watts, J. Mck. and Dunphy, J. E. The Role of the Common Bile Duct to Biliary Dynamics." *Surgery, Gynecology and Obstetrics.* 122:1207–1218, 1966.

185. Westphal, K. "Die Durch Diskinese der Ansfuhrungsguage Bedingten Pankreas-Fermit Shadigungen an den Gallenwegen und der Leber, Klinische und mit Dr. Kuckuck Gemeinsam Durchgefurte Experimentelle Untersuchungen." *Z. Klin. Med.* 109:55–117, 1928.

186. Westphal, K., Gleichman, F., and Mann, W. "Gallenswegsfunktion und Gallensteinleiden. IV. Anatomische Untersuchungen an erkrankten Gallenwagen. *Z. F. Klin. Med.* 115:150–208, 1931.

187. Whitaker, L. R. "The Mechanism of the Gallbladder." *American Journal Physiology.* 78:411–436, 1926.

188. Whitaker, L. R. "The Relation of Biliary Dysfunction to Lithisis." *New York. State Journal Medicine.* 34:221–236, 1934.

189. White, T. T., McAlexander, R. A., and Magee, D. F. "Gastropancreatic Reflux after Various Gastric Operations." *Surgical Forum.* 12:286–288, 1962.

190. Wilkie, A. L. The Bacteriology of Cholecystitis: A Clinical and Experimental Study." *British Journal Surgery.* 15:450–465, 1928.

191. Williams, R. D. and Huang T. T. "New Technique for Experimental, Repeated, Long-Term Measurement of Biliary Pressures." *Surgery.* 65:454–461, 1969.

192. Williams, R. D. and Huang, T. T. "The Effect of Vagotomy on Biliary Pressure." *Surgery.* 66:353–356, 1969.

193. Wingert, W A. and making the, VP G. "Cholelithiasis and Cholecystitis in Childhood." *California M.* 107:26–32, 1967.

194. Wittich, W. V. "Zur Physiologie deer Menschlichen Galle." *Archiv. F. D. Ges. Physiol.* 6:181–184, 1872.

195. Wolfer, G. A. "The Role of Pancreatic Juice in the Production of Gallbladder Disease." *Surgery, Gynecology and Obstetrics.* 53:433–447, 1931.

196. Wolfer, J. A. "Pancreatic Juice as a Factor in the Etiology of Gallbladder Disease." *Surgery.* 1:928–938, 1937.

197. Womack, N. A. and Bricker, E. M. "Pathogenesis of Cholecystitis." *Archives of Surgery.* 44:658–686, 1952.enta

198. Wormsley, K. G., Mahoney, M. F., and Ng, M. "Effects of a Gastrin-Like Pentapeptide (ICI 50, 123) on Stomach and Pancreas." *Lancet I.* 993–996, 1966.

199. Wyatt, A. P. "The Relationship of the Sphincter of Oddi to the Stomach, Duodenum, and Gallbladder." *J. Physiology (London).* 193:225–243, 1967.

200. Wyatt, A. P. "Effect of Gastrectomy on Biliary Dynamics." *Gut.* 10:91–93, 1969.

201. Zajtchuk, R, Amato, J. C., Shoemaker, W. C., and Baker, R. J. "The Relationship between Blood Glucose Levels and External Pancreatic Secretion in Man." *J. Trauma.* 9:629–635, 1969.

202. Zucker, T. F., Newberger, P. G., and Berg, B. E. "The Amylase of serum in Relation to Functional States of the Pancreas." *American Journal Physiology.* 102:209–221, 1932.

Hemorrhoids: Why?

Some important issues are not frequently discussed in polite society; among them is the topic of hemorrhoids! Every species in the animal phylum has its eyes and anus at opposite ends of the carcass—certainly encouraging anal aversion. Thus, there is considerable room for ignorance and misunderstanding of hemorrhoids in the function of the gut—both in health and disease. At the front end, the mouth and the esophagus are derivatives of the skin embryologically. The same is true at the anus and lowermost section of the rectum where the gut and skin fused late in embryonic development. Hemorrhoids are submucosal, longitudinal individual venous columns, extending downward from the mucosal-epithelial junction (about a finger's length up) to the visible external verge of the anus. Hemorrhoids are enclosed within a lacy, loosely attached subcutaneous layer of connective tissue that permits some gliding shift of the overlying mucosa (minimizing static friction during bowel movements). There are three main hemorrhoids, each of which is always located in a predictable distribution around the circle of the anus. With the observer looking upward, these veins are invariably located at three o'clock, seven o'clock, and eleven o'clock; that is, at 120° separation.

We have previously discussed many digestive functions that are carried out by the gut through a combination of cellular, chemical, and mechanical activities. The gut also, from mouth to anus, is a channel for the residence and function of countless trillions (plus/minus 10 trillion?) of bacteria and fungi that may function variably as our allies in digestion or as enemies in disease. Our understanding of these functions is extremely limited.

Another factor of importance that we have largely ignored in the physiology of digestion is that of friction of the gut contents in moving over the mucosal lining of the intestinal tract. In fact, it surely must be the case that every bowel movement we have had since birth has entailed some degree of fecal-cellular friction with the opportunity for micro-inoculation of fecal bacteria. As a result, we develop local resistance (partial immunity) in the perianal and perirectal tissues and, probably, at higher levels.

The issue of friction in soft tissue interfaces has not been studied effectively. Nature has given us excellent protection in joints and their surrounding soft tissues (bursae, gliding surfaces, etc.), but there is inescapable friction (both static and kinetic) when nature has to propel a (variably) firm surface up the pelvic glide to the anus and, with added pressure, open the trap door!

Among the very few studies to be found in the literature relating to these issues is a publication from the Department of Mechanical Engineering, Indian Institute of Technology, Delhi. This was a study of the optimum grasp (minimum friction) for passage of a round bar through a narrow channel. The answer proved to be three contact points around the circumference of a bar, spaced at 120°! How did nature guess it? (A. Subramanian. S. Mukerjee; Proceedings of 11th National Conference on Machines and Mechanisms, December 18–19, 2003; New Delhi. Allied Publishers, XX, p. 781).

Despite nature's precautions, we sometimes develop perianal infections. Most frequently, they develop in a small mucosal recess about a finger's length above the external anus. Usually, these abscesses drain spontaneously and heal. Occasionally, the abscess dissects through the rectal wall into intermuscular spaces within the pelvis then finds its way to drain externally through the perianal skin. This may become a chronic sinus which allows some drainage of feces, bypassing the anal sphincter—the so-called "fistula-in-ano." Other conditions such as chronic inflammatory bowel disease or tuberculosis may cause this condition, adding to the confusion and difficulty in therapy.

This is an ancient problem, still unsolved, and since the days of King Charles V, frequently treated by a so-called seton (adjustable, tied suture loop) to cut gradually through the sinus tract and permit drainage and healing of the tissues behind the suture loop. This is not highly satisfactory treatment—replete with blood, pus, feces, and pain. In my own practice, I passed a grooved instrument through the tract, lifted it, and cut meticulously through the mucosa and underlying muscle identifying and tagging each muscular strand so that it could be matched with its fellow in the wound closure; and then excised the entire fibrotic, infected tract. I then closed this defect loosely but repaired the anal sphincter muscle with precise end-to-end repair of each muscular strand (fascicle). In fact, these wounds all healed with no infections and with satisfactory function of the anal sphincter after recovery. My theory was: An uncontrolled, septic scar through the anal sphincter cannot do more to preserve fecal continence than a cleanly made transverse cut avoiding nerves and with precise repair of every structure in the sphincteric bundle! In cases so treated, there was no damage to the innervation of the anal sphincter.

This approach is not sanctioned by majority surgical opinion. It is. actually, vigorously condemned! The problem of treatment of anal fistula needs to be carefully studied under controlled conditions (I hope, by the Veterans Administration) in order to be generally accepted by the medical profession. Without institutional support and legal protection, any surgeon deviating from the King Charles V barbaric formula in treating anal fistula will be pilloried. The present state of surgical management of anal fistulas is, simply put, one step short of barbarism.

This tragedy is illustrative of what I call the "Semmelweis syndrome"—patient suffering secondary to physicians' traditional dogma and smug ignorance-to commemorate the Viennese obstetrician who was drummed out of the corps for suggesting that a physician should wash his hands before delivering a baby.

There are many other aspects of anorectal physiology that are totally neglected. One of the most remarkable features of rectal function is that the owner is usually able to determine whether the "urge" is due to gas or more substantive pressures. How do we know this? Obviously, the result of highly refined sensory input from the lower rectal mucosa! To this point, a total mystery! Further, this magical talent is not limited to us. If you ever had a dog, you will recall that he would go happily around the house passing flatus without the slightest anxiety. However, when he needed to fertilize the lawn, his face became anxious, and he began to trot or run anxiously toward the door, meanwhile instructing you to open and let him out—*pronto*!

Of course, there are conditions of hemorrhoids that require the attention of a surgeon: notably, pain, bleeding, prolapse, and thrombosis. Proposed methods of treatment are legion and, some, very destructive. Perhaps the least intrusive management is the strangulation by rubber band application over the upper end of the hemorrhoid within the anus. Certainly, the most pernicious treatments are those that advocate division and reanastomosis of the upper anus, which drives a fire-truck over the delicate neuromuscular reflexes that we have just discussed. If hemorrhoidectomy is indeed required, I prefer a vertical submucosal excision with proximal ligation of the vein and loose mucosal closure with fine absorbable sutures. If such a patient awakens from anesthesia, clutching his buttocks with both hands and screaming in pain, his surgeon has done him no great favor. It is a remarkable fact that these procedures are almost never complicated by local infection—a tribute to the local immunity to bacterial infections exhibited by anal tissues which has been discussed earlier.

However, serious infections of the anus and rectum do occur—most frequently among patients with depressed immune resistance from disease or from medications, as from cancer chemotherapy. This condition is called Fournier's gangrene and is frequently fatal. It is of more than passing interest that the first well-described case of this malady was recorded by the Roman historian Josephus and the patient was King Herod the Great. This was the terminal event among such episodes as a visit from wise men from the Orient, the slaughter of the innocents, a high-society strip dance with seven translucent veils, the beheading of John the Baptist and presentation of his head on a platter to the stripper, and the construction of the Temple in Jerusalem. Few patients will meet all the criteria!

Fournier's gangrene is usually an insoluble problem, combining virulent, aggressive, multi=bacterial disease with an already decaying host—not an effective prelude to longevity!

Other Priority Research Problems
Tetanus
The Lurking Serpent

During July 1960, I took my family to Haiti where I spent the month as a volunteer surgeon in the Albert Schweitzer Hospital, Deschapelles. There, I was confronted, face-to-face, by tetanus—an illness forgotten in America since the adoption of universal tetanus toxoid immunization. However, that hospital had an entire ward filled with tetanus patients, mainly infants who had contracted the illness through contamination in the process of severing the umbilical cord. The complexities of treatment had been refined to a degree that remains the current standard of care, including discovery that Valium is an effective (temporary) antidote for the acute neurological agitation of the illness. (Garnier, M. Tetanus. Am. J. Surg. 129:459–463, 1975).

This is an illness caused by an anaerobic bacterium (*Clostridium tetanus*) that inhabits the intestines of animals (including man), usually passing through the gut without causing problems. However, if the bacterium is introduced into the soft tissues through an external wound, its spore (hibernating) phase has the capability of converting to an active metabolic, toxic bacterial pathogen. In this phase, the organism propagates and secretes a toxin that is transmitted through the myelinated sheath of nerves connecting to the brain and by this process disrupts normal nerve impulse transmissions arising within the central nervous system. It is almost universally fatal. There is no known effective treatment to control this process. More attention must be paid to assure universal immunization of infants with tetanus toxoid and to provide booster doses in adult and late life. Further (as is noted in a following section), chronic, low-grade intestinal mucosal tetanus infection may have a role in systemic neurological dysfunction.

Another member of this bacterial family, *Clostridium difficile*, is known to cause severe infections of the intestinal tract. It is readily transmissible by feces, linen, and hand contact. This has become a common scourge in American hospitals and is causing enormous costs for its management and control, as well as the morbidity, prolonged hospitalization, disability, and death in its wake. Fortunately, there is now available an effective *C. difficile* toxoid for prevention of the infection. This disease, in itself, is an imperative challenge to investigate more effectively the bacterial flora of the intestine.

The Enigma
Mental Illness

While I was a medical student, I was introduced to the subject of mental illness during a course conducted at the Boston Psychiatric Hospital. At that time, people who were incapable of caring for themselves were institutionalized for care and control. In the interim since, much has been learned regarding the function of the brain and, to some degree, pharmacological intervention to minimize the destructiveness of these disabilities. However, understanding the cause and treatment of these maladies is still primitive.

During my surgical residency, on a rotation in neurosurgery, I was taken to the McLean Hospital (psychiatric division of the Massachusetts General Hospital) to assist in the operation of prefrontal lobotomy on a violent inmate. Within the span of one hour, we converted a sensate human being to the level of a compliant ape—a shattering emotional experience for me.

I have come to realize that disordered human behavior has been a curse of members of the human race since, at least, biblical days. And yet during a conversation with one of my colleagues, recently retired as a chief psychiatrist at the Massachusetts General Hospital, he affirmed that the cause of schizophrenia is still unknown. This has reinforced my conviction that research to discover this trigger is of the highest importance. As it happens, some related personal experience has emphasized the importance of this issue, and I have come to suspect that both autism and schizophrenia (and, perhaps, senile dementia and other central nervous system derangements) may be related to chronic tetanus infection of the intestinal mucosa. It is my hope that the VA may take leadership in a comprehensive study of the problems of human gut bacteria as sources of human disease.

Letters to Children
The Long Journey

Second cousin Ruth Rodkey Barnard, the family expert on genealogy, did me no favor when she reported, "The Rodkey men all died of lung problems!" Thus, in the late winter of 1988, while I was home suffering from a very severe bout of flu, her dire prediction came to my thought! As it happened, I had been home for three days enduring a severe respiratory infection with heavy cough, chills and fever, and intermittent "lightning" pains through my muscles—a "never event" for me. But instead of recovering, I was steadily worsening. Finally, on the morning of the fourth day, I realized the seriousness of my condition. I telephoned my physician, Dr. Perry J. Culver, a good friend since the day in late September 1939 when I walked into the Harvard Medical School and found him at the front door—a member of the (then) second year class.

"Perry," I said, "I may be in trouble. I have what I think is the flu, and I'm getting progressively worse. The only medication I can think of which might be helpful for a virus would be amantadine. Do you have any suggestions?"

"Yes, that is a good thought, but Flumadine might be better. I will order some for you." I began the medicine in the late afternoon, but I remained extremely uncomfortable with shortness of breath, chills and fever, and intermittent lightning pains in my muscles. I slept fitfully. Suddenly, I was awakened with the sensation of traveling through the darkness at a tremendous rate of speed. Soon, I could see far ahead a faint glow of red light. This enabled me to realize that I was in the right armpit of some very large individual who seemed to be at least four times my own height! This individual was dressed in a black cloak with a hood, and I never saw a face. There was not a sound and not a word was uttered.

We continued our extremely high-speed approach toward the red glow. Finally, as we approached, I saw black flecks moving about against the red glow. Shortly thereafter, to my great shock, I realized that these were humans jumping about, their nude bodies being silhouetted against the red glow! As we approached, I could see the expressions on their faces—all appearing extremely sad and agitated. In particular, I remember a young woman who, after leaping about for an extended time, sat down on a large rock with an elbow on her knee and her chin in her palm with an expression of indescribable sadness.

Suddenly, we were approaching a grayish structure that appeared, perhaps, to have a huge spherical outline, and immediately ahead, a rectangular door opened. As it opened, some of the black figures in the pit jumped, trying to reach into the open threshold, and I was fearful that we might strike them. However, we flew in safely, and the doors were closed behind us.

I found myself in a totally dark area, except that up an incline toward the back, there was an arched doorway through which shone a very bright light—light that shone at approximately twice the intensity of sunlight (or near the intensity of bright sunlight reflected off the side of a white house), very soft and without glare. There were no persons or other figures.

In some way, I found myself turned around to face the wall through which we had come. The door had been closed, and there displayed on the wall was a huge electronic tote board. As I looked, I was immediately aware that each of the squares contained an item that described circumstances of a person whom I had injured or whom I had mistreated in some way—whether intended or unintended. Although many of these were incidents that I had completely forgotten, they were each one immediately recalled to me in excruciating detail, and I felt convicted as to their truth! But, strangely, I felt no sense of condemnation but only a sense of indescribable sadness and remorse.

Then, in some way, I was turned around again and faced toward the glowing arched doorway. We walked up the incline and into the doorway through which I could see many houses with what appeared to be crystalline, translucent walls. No people were evident. There were copper-colored streets that appeared to run to the right for a rather long distance and then curve to the left and were lost to sight. Suddenly, a group of people came walking along the street, appearing to be talking and laughing together. I was unable to identify the features of any face.

Then we were traveling again in the dark. Once more, I was tucked beneath the right arm of my silent companion. We traveled for a huge distance. Suddenly, directly before us appeared the stone spires of a temple (having the appearance of the Mormon temple in Salt Lake City). I was terrified as we slammed up against the stone surface of the tower on our right, but we passed through it without any sensation. Immediately, I saw the workings and significance of the temple! A couple (man and woman) come into the temple. They are met by a female and a male attendant who takes each one aside to a preparation room. There, each one showers and changes into temple clothing. Thereafter, they are brought together and led to a small elevator that takes them to a level midway in the height of the temple. There, they enter a small corridor off which doors open into a small living spaces that overlooks the interior of the temple. These rooms were pointing obliquely toward the front (or altar end) of the sanctuary. These cubicles were so arranged that one could see only forward but not into the adjacent similar cubicles. And none could see into the cubicles across the cavern of the temple. On the bedside table were bottles of perfumes and lotions. There was a small bathroom and living space. Immediately, I realized the significance of this. This was a recapitulation of the Garden of Eden experience of Adam and Eve—a man and woman finding each other in God's presence.

And then I found myself at home in my own bed! After a few days' recovery, I went back to work. I have a friend, also a surgeon, who is a Mormon. When I related this experience to him, he went pale. "Grant, all I can say is, you have had a vision." He would not discuss the matter further.

A few years later, the Mormons built a new temple in Belmont, Massachusetts. During the period after its construction and before its dedication on October 1, 2000, the temple was opened to the public for visitation. When I walked in, I already knew the geography of the place and its functions! Although our tour did not include the cubicles described above, in the main temple chamber, there was a balcony on either side high above the pews. I went to the back of the sanctuary and, from the extreme right corner, peaked up over the balcony on the left side. Just visible was a row of the top edges of door casings extending along the wall!

Our daughter, Cheryl Anne, was asthmatic and at times became quite short of breath. On one occasion, she had a sensation of leaving her body, floating around the ceiling, and watching people struggling to restore her breathing. She told me that she had made a conscious attempt to return to her body and then awoke in her bed.

There are many written and published accounts of so-called out-of-body or near-death experiences. Many of them have similar elements in their descriptions that are analogous to Cherie's and my experiences. A recent case of such an episode in the life of a neurosurgeon (with multiple references) has been published by Eben Alexander, MD. *Proof of Heaven*. Simon and Schuster Paperbacks, 1230 Avenue of the Americas, New York NY, Copyright 2012.

Reflecting on the meaning of such experiences, it has come to me that, throughout the history of mankind, individuals have had many near-death experiences. Falls, concussions, lightning strikes, high-voltage electric shocks, near-drowning, infectious disease crises, cardiac arrests, etc., have taken us to death's door. Many have described their experiences in terms approximating Cherie's and mine. Could these be related to some of the visions of the prophets?

As a potential instance, I quote a description of the experience of Saul of Tarsus, recounted in the Holy Bible, Book of Acts, chapter 9:

> But Saul, yet breathing threatening and slaughter against the disciples of the Lord, went unto the high priest [in Jerusalem] and asked of him letters to Damascus unto the synagogues, that if he found any who were of the Way, whether men or women, he might bring them bound to Jerusalem. And as he journeyed, it came to pass that he drew nigh unto Damascus: and suddenly there shone round about him a light out of heaven: and he fell upon the earth, and heard a voice saying unto him, "Saul, Saul, why persecutest thou me?" And he said, "Who art thou, Lord?" And he said "I am Jesus whom thou persecutest; but rise and enter into the city and it shall be told thee what thou must do." And the men that journeyed with him stood speechless, hearing the voice, but beholding no man. And Saul arose from the earth, and when his eyes were opened he saw nothing; and they led him by the hand and brought him into Damascus. And he was three days without sight, and did neither eat nor drink.

> "Now there was a certain disciple in Damascus named Ananias, and the Lord said to him in a vision, "Ananias!", and he said "Behold, I am here, Lord" And the Lord said unto him, "Arise, and go to the street which is called Strait, and

inquire in the house of Judah for one named Saul, a man of Tarsus: for behold, he prayeth: and he hath seen a man named Ananias coming in, and laying his hands on him, that he might receive his sight." But Ananias answered, "Lord, I have heard from many of this man, how much evil he did to thy saints in Jerusalem: and here he hath authority from the chief priests to bind all that call upon thy name." But the Lord said unto him, "Go thy way: for he is a chosen vessel unto me, to bear my name among the Gentiles and kings, and the children of Israel; for I will show him how many things he must suffer for my name's sake." And Ananias departed, and entered into the house; and laying his hands upon him said, "Brother Saul, the Lord, even Jesus who appeared unto you in the way which thou camest, hath sent me, that thou mayest receive thy sight and be filled with the Holy Spirit." And straightaway there fell from his eyes, as it were, scales, and he received his sight; and he arose and was baptized and he took food and was strengthened. And he was certain days with the disciples that were at Damascus. And straightaway in the synagogues he proclaimed Jesus, that is the Son of God. And all that heard him were amazed, and said: "Is this not he that in Jerusalem made havoc of those that called on this name? And he had come hither for this intent, that he might bring them, bound, before the chief priests. But Saul increased in strength, and confounded the Jews that dwelt in Damascus, proving that this is the Christ.

Was this a near-fatal lightning strike?
Or consider Dante's *Comedia*.

What is the meaning of these experiences? We cannot see through the veil, but we certainly must see significance from the threads of similarity in the lives of so many individuals. I know people who will pass this off as "brain dysfunction due to anoxia and acidosis."

I think the Boss took us by the hand and led us though. We don't run it.

Letters to Children
Saint Barbara

In the mid-1960s, a woman came to our surgical office for treatment of an injury sustained in a fall at work. She was working in the cafeteria of English High School in Dorchester, Massachusetts, and had slipped on a wet floor, sustaining moderate injuries that, in due course, healed satisfactorily. However, the incident began an interaction and devoted friendship that lasted until her death some twenty years later. For reasons which will become obvious to the reader, she was—and forever will be—Saint Barbara.

Saint Barbara was born in the city of Wadowice, Southern Poland, near which the Nazis later constructed the Dachau Concentration Camp for the systematic extermination of Jews. In her youth, she attended a church parish under the tutelage of a young priest named Father Karol Jozef Wojtyla, participating in many of his youth expeditions. Although she never described her home, it was evident from her demeanor, sparkling intellect, and broad education that she had come from a family of significant culture. During her high school years, one of her girlfriends, a Jewess, was seized by the Nazis and thrown into the Dachau Camp. She suffered severely, in part from starvation. Barbara began, surreptitiously, to bring her friend food, which she passed to her through the metal woven fence. The Nazi guards spotted her doing this, captured her, and threw her into the camp. There, she was tattooed with an identification number, starved, given barely enough water to survive, and subjected to surgical experimentation, which included large incisions on her arms that left heavy scars as lifelong reminders. Fortunately, this was near the end of the war, and American soldiers liberated the camp before she died. At that time, Barbara—only eighty pounds in weight—was sitting on a large rock in the yard, perceiving events around her as in a dream. The American soldiers burst into the compound at a time when a large stack of human bodies formed a wall at the foot of a small incline in the yard. When the soldiers saw a pair of legs moving near the bottom of the pile, they went crazy and shot every German in the place, according to Barbara's report.

Somehow, after her liberation, Barbara met a young man (Sigmund), a metalworker, and they were able to immigrate to the United States. Living on a nice street in Dorchester, they had no children, but Barbara was mother to all the children in the neighborhood who could not resist the pungent fragrance of her baking! These happy relationships went on for many years as a generation of "her children" matured. In the interim, Barbara sent frequent food packages and other essential supplies to her aging mother in Poland. Some of these got through despite the hazards of the Iron Curtain, and they played a significant role in her mother's survival.

In October 1979, Pope John Paul II, the former Father Karol Jozef Wojtyla, visited Boston. In preparation for this visit, the Polish community of Boston (among whom we had many patients) unanimously chose Barbara to be his official hostess. This event was a period of great happiness in her life, but it was short-lived. Shortly thereafter, Sigmund was killed. He was seated in the backseat of a car that was rear-ended while stopped at a red light as he was on his way to work.

To conserve her money, Barbara moved to a second-floor apartment in Quincy. There, she made friends with new neighbors and began a pattern of arising at 3:00 AM, going to the local bakery and buying defective packages of pastries (especially doughnuts!) that she brought home to her garage. There, she culled the defective pieces and fed them to the surrounding birds and squirrels, repackaged the remainder, and distributed them to the local firehouse and police headquarters!

But this calm was short-lived. She developed symptoms of abdominal pain that proved to have been caused by cancer of the colon. Operation to remove the tumor was palliative only, as the tumor had already spread widely throughout the abdomen.

Barbara lived the rest of her days in her second floor walk-up apartment, supporting and being supported by her neighbors and her many, many friends.

Saint Barbara, indeed! And preceding your beloved friend Karol Jozef Wojkyla (Pope John Paul II) into sainthood! We thank you for the blessing and lessons of your friendship!

Mr. M

There was a gentleman in Boston's North End, Mr. M, who was a greatly beloved local celebrity. He was a bachelor, having been briefly married then abandoned by his wife. He was renowned as a poet, composing his tender and sentimental poems in Italian. He was known for his gentle and kind disposition. In due course, Mr. M was referred to our office for treatment of hernias.

As it turned out, Mr. M had been in the Italian army in 1915 when he was discovered to have a hernia. He was sent to a military hospital to have the condition repaired. On the day prior to his scheduled procedure, the soldier in the next bed was taken to the operating room for hernia repair, but he never returned to his bed, having died during the procedure! Mr. M, in great alarm, got out of bed, took up his rifle, and went back to the military front to take his chances in combat! He had never, in the interval, seen another doctor until he was referred to our office!

Mr. M wore trousers with the crotch at his knees, and he walked with difficulty. Actually, his scrotum had enlarged down to his knees, his intestines had all slipped down into the scrotum, and his abdominal wall was plastered flatly against his spine. His penis was invisible. This condition was previously known to medicine and is called "loss of abdominal domain." In such a situation, it is impossible to push the intestines back into the abdomen without forcing the diaphragm up into the chest and causing oxygen deprivation and thereby causing death by blocking respiration.

As it happened, one of my associates, Dr. Willard Johnson at the Boston Veteran's Hospital, had recently repaired a large hernia by preliminary injection of air to stretch the abdomen then subsequent hernia repair. It was the only hope for Mr. M.

I began filling the abdominal cavity with injections of air through the abdominal wall at weekly intervals. The abdomen began to expand, and at a surprisingly rapid rate! After six weeks, I repaired the right hernia with no ill effects!

After a recovery period of a month, we began the process again. In another six weeks, we repaired the left hernia, again with no problems. However, after three more months, the right hernia recurred. This was again repaired, and this time with insertion of plastic mesh within the repair. Recovery was uneventful, and Mr. M was cured! The transformation was striking: he could wear normal clothing, he could walk without waddling, and his penis was now visible and functioning. He had no pain.

A few months later, the North End awakened one morning to a grisly discovery. Mr. M had built a huge fire in his stove, burned all his poetry, and committed suicide by slashing himself across the neck with a butcher knife! How could this gentle man have committed such violent acts? His friends were stunned!

My own reflection on these events convinced me that Mr. M, having learned that his fifty plus years of suffering and social isolation had—in the end—been preventable and were now irretrievably lost, simply could not bear the pain!

Mr. M taught me another lesson that I believe is not widely understood, although highly significant. The human abdomen (both male and female) is peculiarly adapted to rapid large changes in capacity to accommodate tumors, ascites, hernias, and pregnancy!

Letters to Children
The Resilience of the Human Soul

In the 1960s, a Polish-American nurse working in the operating room of a neighboring Boston hospital was sent to me for correction of a surgical problem. This was quickly repaired, but not before the establishment of a firm friendship.

Ms. X, married to a Polish army colonel, was working in a Polish military hospital in September 1939 when the Germans invaded. Almost immediately, the Russians rushed in from the east and overran her city. Russian soldiers shot her husband and took her two-year-old baby girl by the feet and swung her head hard against a door casing, killing her on the spot. Ms. X was carried away and sent to Siberia where she was billeted in a lumbering camp in a large forest. The camp was, in fact, a prison camp for women who were being used to fell trees, saw them into logs, and transport them by sled to a sawmill. The women were all billeted in square cabins, each with eight double-deck bunks and with outdoor water spigots and latrines. Lighting in the long winter Siberian nights was by candlelight. Each morning, the women took a horse with sled into the forest, felled and sawed trees all the daylight hours, then returned to their bunks. If any one of the prisoners became too weak to walk, she was simply abandoned to die by the pathway.

As December approached, the women began saving fragments of candle wax to fashion a rosary. They saved bits of food in preparation for a night celebration of Christmas—for which they had prepared the rosary. At the sight of their Christmas devotions, the Russians were furious, kicked in the windows, and placed fire hoses into the cabin, which they filled with water to the sills of the windows, keeping the women imprisoned for twenty-four hours. For most of this time, they had to stand thigh-deep in the frigid water. Mrs. X said that, at the conclusion, her uterus had completely everted from her vagina.

All these events were occurring at the time when Winston Churchill was advocating invasion of Europe from a southern front. He persuaded Russia and the Allies to recruit Polish prisoners in Russia to create a Polish Army to deploy in the Italian invasion. Thus, Mrs. X became a member of this army. In fact, she was in multiple heavy battles, including a three months' stance in the Battle of Monte Castello in 1944–1945.

At the end of the war, in some way, she made her way to the United States and to Boston. Here, she had remarried and had another daughter. Mrs. X was teaching her daughter to be a pianist and, in fact, sent her to the New England Conservatory of Music where she became a very accomplished performer. The daughter met a young man who was a cadet at the Naval Academy in Annapolis, Maryland. The couple was married in the chapel in Annapolis when he graduated!

Faith, hope, love, and the hand of God?

Letters to Children
The Hunchback of Brookline

Among our patients was a middle-aged man with significant spinal deformity. He was a pleasant, courteous, well-spoken gentleman with a story far surpassing in importance his current, mild surgical problems.

In the late 1930s, he lived with his family in southern Germany. On a certain day, while he (aged fourteen) was away from the home, German soldiers came and took away his father and mother. When he arrived home, he was terrified and found a narrow climbing space to the attic where he took up residence. Later in the day, German soldiers came to stay overnight and to make this their barracks.

After they left on the next morning, Mr. J crawled down and left the house to find food and water. He met another child, a ten-year-old daughter of neighbors whose parents also had been abducted on the previous day. The two returned to Mr. J's house and attic sanctuary. They continued living in this attic, crouched in the corner, with daytime foraging for food and avoiding the soldiers for the remainder of the war—some three years. According to Mr. J, it was during this period that he developed his "hunchback" in the cramped living quarters under the eaves. Finally, they were liberated when American troops came through the area at the conclusion of the war. The two teenagers had no other friends and no other family, so as soon as possible, they married.

Through the assistance of American soldiers who rescued them, they found an opportunity to immigrate to America thence to Boston. Living a quiet life—both working as clerks—they continued their love and support of each other.

The enduring, transforming, saving power of love?

Letters to Children
First-Rate Mind / Fourth-Rate Body

During the early years of my practice, a young man came to see me regarding joint pains in his legs. He related the story that, during World War II, he had been drafted and given a preinduction physical examination. He had been rejected for military service at a time when the popular comment was "if you are warm, you are in!" In fact, the examining physician had told him, "You have a first-rate mind in a fourth-rate body!"

Not long thereafter, I got an emergency telephone call at 5:00 AM from his sisters with a request that I come immediately to their apartment in Dorchester to care for their brother! I hurried out, found the apartment, and went in to find the brother totally paralyzed. The condition had come on during his sleep; he was unable to speak and unable to move any part of his body except his eyelids! This was certainly new to me!

With the help of his sisters, I loaded the man into my car and drove him to the emergency department at the Massachusetts General Hospital. There, he was admitted to the neurology service, and he came under the care of one of my medical school classmates, Dr. Edward Pierson Richardson, a neuro-pathologist. As it turned out, this is a recognized illness related to some vascular anomaly within the brain that causes a calamitous stroke and is not treatable. After a few weeks, the patient was transferred to a chronic hospital—the Lemuel Shattuck Hospital in Jamaica Plain. There, he was dutifully and lovingly cared for by his sisters, serving in tandem. Each of these individuals was very creative, and very soon, they developed an alphabet in which the letters were defined by the numbers of eye blinks! This was incredibly cumbersome, but these were all highly intelligent individuals, and they quickly mastered the ability to communicate! In fact, the sisters began documenting the subjective reactions of this patient—now totally separated from the ability to communicate with fellow human beings and totally unable to move any part of his body! This went on until the patient succumbed approximately two years later.

These two devoted sisters continued to support themselves and to live together over the course of more than half a century! As recently as 2015, I met them in the corridor of the West Roxbury Veterans Administration Hospital. We had no difficulty in recognizing each other!

Letters to Children
Tough Old Bird

In the early 1980s, we were sent a patient from Fall River who had colon cancer. This lady was in her mid-eighties, and the question had been posed as to whether she should have an operation. She arrived with her daughter at the duly appointed time. I interviewed and examined her carefully. She was mildly obese and had hypertension that had been treated but had never had heart failure. All in all, she was pretty active physically and had a very positive outlook on life.

After the examination, I sat with the patient and her daughter and discussed the issues. In the end, I told her that I would recommend the operation. I felt that she had a reasonable chance to survive and be well and that she seemed to me to be a tough old bird.

The patient did have a colon resection with the removal of all disease, an uncomplicated recovery, and at least ten years of survival. Each year at Christmas, we received a Christmas card from Fall River with no message but a signature: Tough Old Bird!

Letters to Children
The Most Important Thing

Early one evening in December 2015, I had a telephone call from a gentleman with a weak, quaking voice: "Is this Grant Rodkey?"

"Yes."

"This is Cliff Straehley—do you remember me?"

"Yes, of course, Cliff. Where are you?"

Dr. Straehley had been a surgical resident at the Massachusetts General Hospital when I returned from military service in 1948. In fact, he was one grade behind me, and I became his instructor. He completed his residency in thoracic surgery and thereafter had gone to Honolulu to open a pioneering practice there in that specialty. He had an outstanding career of leadership and service, but he and I never saw each other after he had completed his training.

"Do you remember the most important thing you ever taught me?"

"No, Cliff, what was it?"

"We were sitting together in the ward office on the sixth floor of the White Building when you asked me, 'What is the most important thing a surgeon can give to his patient?' I responded with the usual things: inquire as to his chief complaint, present illness, past history, family history, prior injuries, and operations; do a complete physical examination; and do the appropriate laboratory studies. You said, 'Yes, Cliff, those are all very important. But I believe that the most important thing a surgeon can give his patient is love.' All my life I have remembered that advice and tried to live up to it. I have now retired and am living in Walnut Creek California. I am writing a few reminiscences which I want to send you."

The Shattuck-Richardson Family of Boston

When I arrived in Boston in September 1939, I immediately met this family in the persons of my classmate Edward Peirson Richardson Jr. and his uncle—our professor of tropical medicine—Dr. George Cheever Shattuck. Upon graduation from medical school in 1943, Peirson was appointed as intern in Neurology at the Massachusetts General Hospital, and he remained with that institution in the Department of Neuropathology for the remainder of his life, accomplishing a brilliant career in patient care and neurological research.

Peirson's uncle, Dr. George Cheever Shattuck, taught us the rudiments of tropical medicine, which stood me in very good stead in China in respect both to personal hygiene and the care of our patients. In later years, Dr. George Shattuck wrote *A Memoir of Frederick Cheever Shattuck, MD*, about his father. Generously, he sent me a copy of his book. In responding, I commented how interesting had been his father's graduation address to the Yale Medical School Graduating Class of 1907. This treatise caught the essence of the spirit of medicine and an insight into the knowledge and mores of the profession at that time. I commented that this would be instructive reading for current medical trainees. To my astonishment, shortly thereafter, a truck pulled up in front of my office and unloaded cardboard cases containing twenty-thousand copies of this treatise! There was an attached message: "It is now up to you to deliver these!"

This occurred in 1969, and I am just completing this assignment, having only a few remaining copies!

With the help of medical students, surgical residents, the Massachusetts Medical Society, and other friends, the task is now finished! I shall extend the influence of Dr. Frederick Cheever Shattuck by inserting the content of his short treatise into the manuscript of *Our Fated Century*: "The Science and Art of Medicine in Some of Their Aspects" by Frederick Cheever Shattuck, MD.

These brief comments barely touch the surface of the contributions of the Shattucks of Boston—still continuing into successive generations.

Letters to Children
The Man Who Raised Up Hell in Washington and Got Us Our Hospital!

In the fall of 1967, the board of directors of Blue Shield of Massachusetts held a "retreat meeting" in Chatham Bars Inn, Cape Cod. The business of the meeting is long forgotten, but the after-dinner speaker memory is immortal. He was Dr. William Gifford of Colebrook, New Hampshire. His title was "How a Woman Is Like an Airplane"! He began, "To start, you have to give her propeller a spin!" And from that point for twenty minutes, the audience was roaring in laughter! Thus began a treasured friendship!

Dr. Gifford was a general surgeon in Colebrook, New Hampshire, as well as a licensed pilot. His wife, also a general surgeon, participated in the practice. Their practice covered the area of Northern Vermont, Northern New Hampshire, Northern Maine, and the adjacent areas of Southern Canada; and their airports were pastures, fields, or highways. Their home base was a small hospital in Colebrook, then named the Connecticut Valley Hospital. On one of my visits to Colebrook, one of his fellow townsmen told me, "He was the man who went down to Washington, raised up hell, and got us our hospital!"

The Giffords were leaders not only in their local community, but they were active members in the New Hampshire Medical Society and served in the House of Delegates of the American Medical Association. In these roles, they were articulate advocates for the medical needs of rural and isolated communities throughout the United States and Canada. It is true that Dr. Gifford went to Washington, DC, shook up the Congress and the administration, and got the funding to build the Valley Hospital in Colebrook—a man of magic, madness, and miracles!

But the most remarkable thing about Dr. Gifford is that he should never have been born!

Dr. Gifford's father was a Scotch immigrant who settled in Manchester, New Hampshire, and found work as a cook. When he became secure in his job as planned, he sent word (and money) to his fiancée in Scotland to come and join him.

She was scheduled to come aboard the HMS *Titanic*. Communications thereafter were both confused and slow, and Mr. Gifford assumed that his fiancée went down with the ship. However, about three weeks later, she showed up in Manchester! Her father, a longtime dry-dock construction worker, when he heard of the plan, said, "No, you can't go on the *Titanic*. It will be a maiden voyage, and there will be a lot of drinking. Ship management and pilot watches may be sloppy, and it may be a risky voyage. You can go on any other ship, but you can't go on the *Titanic*!" So his daughter waited until the next sailing to New York and arrived safe and sound!

Experience counts!

Henry Knowles Beecher, MD
Anesthetist, Visionary, Enigma, Friend
A Salute

Dr. Henry K. Beecher was, throughout his entire life in Boston, a highly successful, emotionally driven, highly sensitive, creative, elusive, generous, brilliant friend! Many publications have touched on the various facets of his character. I shall cite only one article, which will serve as an entrée to the extensive publications that deal with his life and work, namely, "Henry K. Beecher—the Introduction of Anesthesia into the University" written by Dr. J. S. Gravenstein; Anesthesiology 1998, 88. 245–253.

The only justification for making additional comments is that my own perspective adds some insight not previously expressed.

As has been well-documented, Dr. Beecher had changed his entire name to erase all connection with his Kansas family before arriving at the Harvard Medical School in 1924. There is no information regarding the motivation for this action, but a substantial degree of personal struggle during his youth seems a reasonable insight. He was accepted among a group of eight Harvard Medical School seniors for appointment to a surgical internship at the Massachusetts General Hospital in 1928. In the spring of 1929, Dr. Edward Churchill, chief of surgery in MGH, arranged for Dr. Beecher to spend a year in England studying anesthesia with the plan of returning to MGH thereafter. This occurred, and in 1933, Dr. Beecher was made chief of anesthesia at MGH, which then had three operating suites: the Phillips House and the Baker Memorial (private services), and (1939) White 3 (for ward service patients). The White Building also housed the offices and laboratories of the anesthesia service.

In the spring of 1941, Dr. Beecher advertised for a medical student to work in his department during the summer of that year. I applied, went for an interview with Dr. Beecher, and was hired on the spot!

160

In retrospect, I think Dr. Beecher saw in me a fellow Westerner and he suspected that I was having a tough climb. As a matter of fact, during my time working during the summer months, I never saw Dr. Beecher in the operating room! I suspect that he was giving anesthesia in the Phillips House and working in his laboratory.

The next time I interacted with Dr. Beecher was during my own interview for appointment as a surgical intern at MGH in 1943. Dr. Beecher was one my interviewers, seated among other staff members, smiling and encouraging me during stressful discussion. I always felt that he had a significant role in my staff selection. During my residency years, we had occasional chance meetings—always cordial and encouraging but brief.

After I began to assist Dr. Arthur W. Allen (in 1950), I worked frequently with Dr. Beecher in the Phillips House operating room. I learned that Dr. Beecher had a very difficult time in placing an intravenous needle for fluid infusion during the operative session—an essential for most procedures. In fact, he became quite nervous and obviously dreaded the ordeal. To make matters worse, some of his colleagues and some of the surgeons involved were not sympathetic. There was often an air of bemusement or impatience in the audience. I felt very sad for Dr. Beecher although I did not, then, understand the issue. In later years, working with many surgical residents, I found two or three individuals with the same deficiency! While working with them, I discovered that each of them had some deficiency in binocular vision, which is an essential ability in order to have depth perception. Obviously, this was the cause of Dr. Beecher's problem—although no one recognized it. I suspect that Dr. Edward Churchill had identified his ineptitude with manual dexterity (although, not its cause) and for that reason encouraged his move into anesthesia as a specialty. Wise and discerning chief!

During World War II, Dr. Beecher went to the Mediterranean theater of war along with Dr. Churchill. There, of course, he heard of the atrocities of Auschwitz and Dachau and later was familiar with some of the international tribunals on war atrocities. He also had, I believe, some misgivings about universal assurance of high quality in civilian surgical innovations. With these issues in mind and an awareness of the delicacy of political relationships in 1955, Dr. Beecher encouraged the establishment of and suggested the first chairman of the Massachusetts General Hospital Committee on ethics, Bishop Henry Knox Sherrill. These efforts have greatly heightened the safety of patient care during surgical intervention.

During the decades of the 1980s and 1990s, there was an almost universal switch from scalpel to electrosurgical incisions. I never switched because I believed (and still believe) that a sharp steel scalpel (no. 22 Bard Parker blade) cut tissues with less trauma than electrocautery. Because I did not use cautery, the anesthesia department used my cases to train all their staff in the use of ether inhalation anesthesia and on how to avoid the risk of explosion and fire. Thus, I became acquainted with each of the staff who came aboard. I treasured having the opportunity to know these dedicated young professionals.

I have always felt that there is an inherent, sacred partnership between the surgeon and the anesthesiologist to encourage the spirit and to preserve the life and safety of our patient—both in the conduct of operative procedures and in the required research and development to carry forth this safe transit. Each of us shares the opportunity, responsibility, and the honor of protecting this temporary suspension of sensate life. There is no more profound responsibility and trust in human society.

Dr. Henry K. Beecher and his staff have greatly enriched and supported my own life and the lives of our many mutual patients. Not bad for a man who had unsuspected faulty visual depth perception!

William Harvey Frazier, MD
for Whom
There Is No Adequate Description

During our first year at the Harvard Medical School in 1939, I became aware of Harvey Frazier from Spokane who had graduated from Whitman College in Walla Walla, Washington. Although we were both living in Vanderbilt Hall, the Harvard Medical School dormitory (I on the sixth floor, north wing, he on the third floor, south wing), we were both too overwhelmed by registration, equipment and textbook purchases, orientation, and early classes to seek each other out for introductions. Finally, in midterm, on a Sunday afternoon, I searched out his room and rapped on his door. He was living with another classmate, Don Enterline, of Casper, Wyoming. After initial introductions, we all got down to urgent business, memorizing the anatomy lesson for class on the following day! This was orientation-immersion season for more subject matter than I care to recall! We quickly developed a system of rotational quiz-answer routine that identified our weaknesses and encouraged our strengths! Besides, Harvey and Don had a nice radio/phonograph and cordial spirits that encouraged repetitive sessions!

I learned that Harvey was engaged to a young woman in Spokane who was ill. Not too many weeks into our friendship, she died. Harvey went home for her funeral and, in addition to the stress of grief and sorrow, he lost several important steps in our class and study routine. Don and I worked to help him regain his footing—all in all, a very stressful time.

In February 1940, Harvey and another classmate, Roger Morrison, introduced me to a Wellesley College student named Dorothea Smith. Dorothea had been born in Nanking, China, of missionary parents who were now back on station so that Dorothea was living in one of the Wellesley dormitories. This connection locked! In 1941, when Dorothea took her master's degree at Columbia University, her parents were imprisoned in a Japanese Concentration Camp in Shanghai. She had no close relatives here and no money, so we got married!

Following graduation from medical school in April 1943, Harvey Frazier took an internship and residency in obstetrics at the Hartford Hospital. Thereafter, he and Don Enterline formed a partnership for the practice of obstetrics in Spokane, Washington, which arrangement continued very happily until Don's premature death about ten years later. Meantime, Harvey had married Amy and begun a family of his own.

Harvey's father died of a coronary attack while we were still in medical school, and his mother continued to live in the family home in northwest Spokane. However, the time came when she felt she must go into a nursing facility. According to a prior arrangement, she and Harvey took a day to survey potential places for her to move.

At the end of the day, she said to Harvey, "Would you mind stopping by the Devonport Hotel for a few minutes?"

"Of course!"

They drew up to the curb, and Mrs. Frazier debarked and went in. She was so slow in returning that Harvey finally got impatient, parked his car, and went into the hotel lobby to locate her. To his amazement, he found that she had checked herself in and was already upstairs in her room! In fact, Ms. Frazier never moved out of that room during the remainder of her life! Each day, she came down to the balcony overlooking the spacious and grand hotel lobby to enjoy sporadic conversations, watch the world go by, and radiate good cheer to fellow travelers on the road of Life!

Meantime, Harvey developed a highly successful obstetrical practice. This involved working at the Deaconess and at the Sacred Heart Hospitals in Spokane and training residents in obstetrics from the University of Washington in Seattle who were sent over on rotational assignments to train with Harvey. Elsewhere in this document, I have spoken of Harvey's help in the adoption of our own two children. His professional skill and his spiritual sensitivity made an ideal combination for appreciating the value of and encouraging the beauty of life at its inception.

Harvey himself had almost a child-like appreciation of joy and happiness. He was completely colorblind and, without the slightest self-consciousness, wore a green felt hat throughout the four years of medical school, thinking that it was brown! Yet in the course of his obstetrical practice in dealing with anxious mothers and small children, he became a vivid and prolific painter—an art achieved by buying numbered paint tubes with the colors identified on the labels by digital numbers. He combined this talent with "prayer prescriptions" in ways to give assurance and comfort, hope and blessing to hundreds of families. In the end, he actually compiled two little printed documents, summarizing these comments, which I will have reprinted as a part of this memoir (see scripture prescriptions).

Harvey and Amy are now both gone, but their family flourishes and their friends, patients, and beneficiaries abound and grow. Those of us who were privileged to know and love them are forever indebted to them for their continuing inspiration and hope.

The Century of Death?
Sadly: The Winner is Dubious

From my early childhood, I began to hear from our Norwegian neighbors rumors of persecution and death in Russian territories in the Arctic—an area abutted by both nations. Although I did not understand, there was an air of anxiety and foreboding. Later, in the University of Idaho Library, I learned more about the Bolshevik Revolution, Russia's genocide, the unrest and conflict in Germany, the invasion of China by Japan, and the deadly slaughter of World War II involving all Europe, Africa, Russia, China, Japan, Vietnam, and Burma. Cambodia (with the final stroke of atomic weapons) outpaces numbers and human comprehension!

But hold on to your hat! When one digs a little deeper, the world is covered by shallow graves! A convenient starting point is Ghengis Khan (1162–1227). With his Mongolian horse cavalry, he conquered Northern Asia from Japan in the east to Northern Europe in the west and to India in the south—reputed to have been an area twelve million square miles and the largest empire in the history of the world! Quite evidently, he spent little time negotiating with the local inhabitants! Body counts were not then in fashion. It is astounding to find the assertion that one in two hundred males alive today is a genetic descendant of the Great Khan!

However, the more doleful news is that genocide was not an invention of Ghengis Khan. In 1991, in the Eastern Alps on the border between Austria and Italy, out from the depths of the "eternal" glacier, protruded a man's torso! The gentleman was carefully freed from the ice and has been preserved by refrigeration since. An astonishing lode of information has emerged from this investigation of which a few points will be mentioned here. The iceman died approximately 5,300 years ago (3,108–3,553 BC). He was carrying a bow, a partially filled quiver of (fourteen) arrows and a flint-bladed knife. The cause of his death was an arrow imbedded in his left axillary artery followed by fatal hemorrhage.

Thus, it is a certainty that man is not in a phase of repentance and reform! The instinct to kill remains, and the tools have escalated. Our atomic weapons are now capable of decimating (perhaps) half a continent with a single blast! And it will happen. Nothing in the evolution of man's heart and mind has changed this murderous, ruthless, destructive, selfish genetic heritage. The question is "Will *our* fated century be humanity's last?"

The Sexual Revolution
Contraception, Abortion, Infanticide,
Slipping the Bonds of Genetically Defined Sex

My years at the Harvard Medical School were a time of intensive study of human sexual physiology. Male and female hormones had been discovered, the physiology of menstruation, insemination, embryonic fetal development, and pediatric care were under intense study. Professors of physiology, biochemistry, pathology, and obstetrics all played contributing roles. Issues of population control were not then discussed, although within the next twenty years, the threat of overpopulation became a political issue.

In addition, religious and morality issues were soon hotly debated—with emphasis on the role of progesterone as an inhibitor of ovulation—thus simulating an extension of the rhythm method of natural contraception. The most renowned obstetrician / scientific investigator in this field was himself a prominent Catholic and a Harvard University professor—Dr. John Rock (see *The Pill, John Rock, and the Church*, Loretta McLaughlin, Little Brown & Company Boston, Toronto, 1982). However intense may have been the theological discussion, women themselves were quick to seize upon the opportunity for new freedom from uncontrolled maternity. *Our Bodies Ourselves*, 1969, appeared as an almost giddy "declaration of independence" from home drudgery. The fever spread rapidly, perhaps culminating in the wild excesses of Woodstock (Woodstock Music Festival, August 5–17, 1969, White Lake, New York, 400,000 attendees). Subsequently, there have developed worldwide epidemics of hepatitis B and C and HIV/AIDS—all diseases transmitted by oral or sexual contact and incurable. Currently, Ebola and Zeta are emerging as additional virus risks for sexual transmission.

Whether related or not, this same time interval has witnessed a marked change in social attitudes toward gender identity. Homosexuality has—at least from biblical times—been a long-standing element of human society. It is still not understood. However, during the past half-century, there has been greatly increased social/political pressure for recognition, acceptance, and tolerance of this condition as a variant of normal. A new feature of current social pressure by affected members of this social group is the demand for access to sex-changing plastic surgical rearrangements; that is, sex-changing modifications of one's natural-born genitalia. This idea is strongly promoted as a civil right owned by affected individuals and to be provided to them by the civil society. Further, it is

165

now stridently advocated that individuals who have had mechanical alterations of natural anatomy have thereby acquired the right to utilize any toilet or recreational water facilities regardless of the gender posting of such areas. This assertion seems overreaching, arrogant, and selfish to many of the citizens who are called upon to bear the economic, convenience, and social costs of such arrangements. Thus, social attitudes relating to homosexuality are sharply polarized at the close of our fated century.

Secular government imposition of population control policies has been carried out in China for approximately fifty years—the one-child policy. Both the effectiveness and the effects of this policy are controversial and confusing. As of 2015, China seems to be rescinding this policy. Objective evaluation of its effectiveness is not available and may not be forthcoming. Meantime, many Chinese baby girls were adopted by American parents and are growing up to become highly productive American citizens! Human sexual instincts are not readily modified!

Letters to Children
Medical Marijuana

That which has been is that which will be; and that which has been done is that which will be done and there is no new thing under the sun.
—Ecclesiastes 1:9

It may be the most shocking discovery in life to realize that I, personally, am responsible for my own attitude and behavior! It is human nature to seek a scapegoat—even Adam tried to pass the buck to Eve. But the Boss was not buying it. Adam was booted out of Eden and told to get to work! How is this relevant?

When I was a boy, I read the shocking story of hashish users in North Africa who were beheaded because they were considered to be unreliable or dangerous. However, I had my own experience that gave some insight: Old Charlie, our loco horse whose behavior was always erratic and potentially dangerous. Eastern Colorado prairies grow several varieties of plants (collectively called loco weed) that cause physical and mental changes that render an animal permanently unpredictable and unreliable in behavior. (Subsequent postmortem studies have shown organic brain damage.) The point is, these chemicals are closely related to marijuana. And Muslims still prohibit their use.

Addictive drugs and deranged human behavior are not new issues. Perhaps one of the most dramatic, illustrative experiences was that of the Chinese Ching

Dynasty (1644–1912), which was brought down in the nineteenth century by morphine and its associated behavioral and fiscal disorders. Drug cartels are not an invention of the twenty-first century.

In the United States, during the latter half of the twentieth century, American psycho-scientists (led by Timothy Leary, a Harvard University professor) have advocated the unrestricted use of psychoactive drugs, extolling their heightened insight into truth and beauty. Objectively, their disciples may actually sink into apathy and reduced capabilities.

A "Systematic Review: Efficacy and Safety of Medical Marijuana in Selected Neurologic Disorders" by Barbara S. Koppel, MD, FAAN, et al. (*Neurology*, 2014 Apr 29:82 [17]: 1556–1563), carried out a long-term study (1948–2013) for efficacy of marijuana treatment of multiple sclerosis, epilepsy, and movement disorders. This huge study had very complex analyses with extremely confusing conclusions amounting mostly to borderline or no clear-cut improvement. No reference to brain morphologic changes was made in this report.

However, for the sake of my younger readers, there should be no confusion about the "Old Charlie" effect on the brains of humans. There is a huge collection of literature ranging in the maybe or "I don't know" territory. But the definitive observation is "What happens organically to the brain as the result of marijuana smoking?"

There are enough studies to be dizzying, but the important studies are those that study the brain itself—fortunately, now possible by imaging and not only by autopsy. Marijuana smoking causes smaller brain development, proportionately more white matter (telephone lines) and less gray matter (working brain cells). Moreover, younger age of marijuana usage was associated with smaller height and weight, especially among males.

Life is an uncertain venture. Each one of us comes with a distinct and different collection of abilities, strengths, potentials, and weaknesses. And as we mature, each of us becomes accountable for his/her own choices and behavior. Each one of us, wherever we find ourselves on that scale of opportunity, has the responsibility to use our talents in the most constructive way which we can for the benefit of ourselves and our fellow travelers. There is a concomitant responsibility to preserve, insofar as possible, our physical assets.

> Know ye not that your body is a Temple of the Holy Spirit which is in you, which you have from God, and ye are not your own? For ye were bought with a price: glorify God, therefore in your body.
>
> —Corinthians 6:19–20

Therefore, my younger brothers and sisters, let us rejoice and give thanks for the assets and opportunities which we have been given in life and to strive earnestly to pass along to others—especially to our successors in life—our hope for those whose lives extend beyond our fated century.

Slow Onset Armageddon?
How Do We Know?

We live amidst great cacophony concerning the warming of the earth secondary to man's slothful waste and self-indulgence. Personal reputations and fortunes are gambled, national governments and international convocations discuss strategies to control this runaway conflagration, and policies are projected to restrict energy utilization—all in confident assurance that these measures may reverse nature's perverse tide!

Few bother to remember that we are only eight thousand years out from the last ice age and, apparently, still in the warming phase. It is only during the past fifty years that man has had the opportunity to see the surface of the earth from outer space—our first inspection sortie since the big bang—approximately 13,799 billion years ago. Why have we been so neglectful of our creation?

On February 12, 2016, astronomers detected the sound of two black holes colliding approximately 13.7 billion years ago—an event noted and the arrival of the sound of impact predicted by Albert Einstein just one hundred years ago. How did he know? Why have the rest of us been so insensitive as to have been surprised by this announcement?

Another relevant topic that does not seem to have been a subject of consideration at our international climate conferences is that of polar shift. Credible scientific view is that several past changes of the Earth's magnetic polar orientation have occurred. It is believed that the last polar shift may have occurred approximately eight hundred thousand years ago, and that such polar reversals may occur approximately at two-hundred- to three-hundred-thousand-year intervals. It is also known that there is slow, incremental creep in the polar magnetic field that is continuous. Shouldn't the United Nations aid us in planning to reverse this process? The business of prophesy is tricky! Its redeeming feature is that the prophet is often, during his lifetime, shielded from disgrace! The fallacy in his predictions has not yet been made manifest!

This is the confused state of mind with which I consider our fated century, and the kaleidoscopic changes that it has brought to each of us. I have reread the book of Revelation, written by the Apostle John near the close of the first century AD. As in previous readings, I find it very confusing, but this time, after my own long journey, a more plausible, the characters seem more familiar! And this time, I realized that we have a powerful interpreter: George Frederick Handel (1685–1785), composer of the *Messiah*. This articulate masterpiece was composed in London during the period August 22 to September 14, 1741—a scant three weeks. It is, in effect, an abbreviated solo and choral summary of the prophecies concerning Christ, his life, crucifixion, resurrection, and his promised return to rule his kingdom without end—the prophecies of Isaiah and a summary of the New Testament including Revelation! Scant wonder that this inspired interpretation, at its first performance, prompted King George II to rise spontaneously and stand during the choral presentation of the divine Hallelujah Chorus to acknowledge his subservience to the King of Kings!

What has this to do with climate change? Perhaps the best answer for this question and for other enigmas of our fated century is to be found in the lines of the bass soloist in the Messiah: "And who shall abide the Day of His Coming? And who shall stand when He appeareth?" We don't know the answers, but we do know that we humans are not running the progress of events unfolding in our world or in the remainder of our entire universe-remarkable as our talents may be!

To return to Handel's divine insight: "And He shall reign forever and ever! King of Kings and Lord of Lords! King of Kings and Lord of Lords! King of Kings and *Lord of Lords*!"

Amen, amen, *a . . . men*!

Letters to Children
Saint John's Book of *Revelation*

During my lifetime, I have read the book of Revelation a few times but always with a sense of confusion and lack of comprehension. However, in writing *Our Fated Century*, I have reread the book, but this time, I discovered an interpreter!

During my years as a student at Whitworth College (1937–1939), I sang in the college chorus where we learned thoroughly and sang frequently George Frederick Handel's *Messiah*. To this moment, I can hear Lowell Poore's beautiful tenor voice in my right ear singing, "For the glory, the glory of the Lord shall be revealed!" But despite close familiarity with and love of this great classic of musical art, its deeper significance had not gripped my heart until this, my ninety-ninth year! The *Messiah*, in its entirety, is an interpretation/translation of the vision and prophecies of Hebrew leaders since King Solomon, combined with the vision of the Apostle John in approximately AD 60–70. Thus, the *Messiah* is a translation key for the book of Revelation!

With this concept in mind, it is of interest to review the circumstances of George Frederick Handel's composition of the *Messiah*. A German by birth into a religious household, the son of a surgeon, he was living in London. There, he found a congenial market for his musical compositions and was able to support himself with a degree of opulence. He remained a bachelor throughout life and was notable for his charity toward abandoned, homeless children. He was a member and congregant at Saint George's Hanover Square Church.

But none of this knowledge prepares us to understand the prodigious—bordering on incomprehensible—composition of this entire work within a three-and-a-half week span in August/September 1741. When one perceives that this effort also included the conceptualization of the entire messianic message from Solomon to Saint John, there are no words to encompass the wonder! How humbly we must stand and bow during the overpowering ascendance of the indescribable Hallelujah chorus of affirmation, adoration, and eternity!

Amen, amen, amen, amen!

And so?

At the end of our fated century, the world seems to be in utter turmoil with its inhabitants bound on self-destruction—this time having the means to accomplish it. Yet if we slaughter each other, if we blast ourselves with nuclear bombs, if we have earthquakes, fire, and floods, if unimagined destruction overtakes us, there still be left a remnant from which our Lord, in his might, may create a new heaven and a new Earth. We do not run the program!

Amen, amen, amen!
Grant V. Rodkey, MD
September 24, 2016

The Gift

A book for children
of all ages

In this book

I have asked THE LORD

to please enable me

to write about

HIS LOVE

for children

--all children

...of all ages

To do this
I have asked
a friend of mine
to help me —

Hello

My Name

is Zeke

Well, really my Name is Ezekiel
    That's what Grandpa
    Harvey calls me.
He named me after a prophet
    who lived a long
    time ago — He
wrote a whole Book in THE Bible —

    — Ezekiel that is.

Who is Grandpa Harvey?
Well, he is someone I Know
I will tell you more about
him later—

You know.

I have some great
things to tell you — that is
why I wrote this little book
for you.
(Actually Grandpa Harvey
helped me a little bit)

You have seen my picture.
Do you have a picture you can
send to me?
I would like that!

Right now I want
you to tell me about
yourself—

How old are you?
Where do you live?
Stuff like that—because
I really want to get to know
YOU!

You can write to me—I'll give
you Grandpa Harvey's address.
He will read your letter to me—
I don't read very well yet.
     Are you a good reader?

How about Spelling?
I have the "dog gone dest"
time with spelling!

Grandpa Harrey
reads to me.
Tell me, who reads to you?

Or are you so almost grown up

that you do your own
reading
and writing
and spelling !

NOW THAT WE HAVE REALLY

MET - I WOULD LIKE FOR YOU

TO MEET SOME OF MY FRIENDS.

MY FIRST FRIEND HAS A

FUNNY NAME - -

HIS NAME IS CREAM PUFF!

HE DOESN'T REALLY LIKE THAT

NAME -

BUT YOU SEE HE EARNED IT - -

HIS REAL NAME IS PETER, HE

IS A NICE CAT - BUT PETER

RUSHES INTO THINGS.

SOMETIMES

PETER ACTS FIRST

AND

THINKS LATER!

that is why Peter looks
So unhappy in this picture.

PETER LIKES MILK -

   AND HE REALLY, REALLY

      LOVES CREAM!

ON THE FARM WHERE HE LIVES

HE ALWAYS MANAGES TO BE ON HAND

WHEN IT IS MILKING TIME -

<u>AND</u> WHEN THE CREAM IS SEPARATED

FROM THE MILK TO MAKE BUTTER -

PETER IS <u>ALWAYS</u> WATCHING - - -

IT IS NOT A MODERN DAIRY FARM

WHERE PETER LIVES. IT IS JUST A

SMALL FARM WITH ONLY A FEW MILK

COWS. AS I SAID PETER IS ALWAYS ON

HAND AT MILKING TIME AND WHEN THE

CREAM IS SEPARATED FROM THE MILK TO MAKE

BUTTER -

   SOMETIMES - -

      SOMETIMES - - PETER GETS INTO

      BIG TROUBLE!

YOU SEE, HE HAD FOUND THIS BOWL OF CREAM

WHICH HAD BEEN PUT ASIDE TO BE USED FOR

REAL THICK WHIPPED CREAM FOR THE

PUMPKIN PIE ON THANKSGIVING DAY!

    WOW!

        WAS EVERYBODY UPSET WITH HIM!

AND AS IF THAT WERE NOT ENOUGH,

SOMEONE TOOK HIS PICTURE

    WITH A POLAROID CAMERA -

WOULDN'T YOU BE UPSET IF SOMEONE

    TOOK YOUR PICTURE

WHEN YOU WERE DOING SOMETHING YOU

    WERE NOT SUPPOSED TO DO?

Peter had been told many
times that it was only
when the milk or cream
was in his very own bowl
that it
was for him.

AND
then
Some one who saw his
picture and saw the cream
on his whiskers
called him "CREAM PUFF"!

Now I want you to meet
            Hildegard -- oophs
I mean Hildegaard -- (two a letters
in her name)
SHE REALLY HAS CLASS!

This picture of her was painted

when she still lived in the sea.

You know "the fish of the sea"

    like it says in the

        8th Psalm.

Why don't you read Psalm 8 in your

Bible now, or ask that it be read

        to you.

Grandpa Harvey reads the Bible.

The 8th Psalm is one of his favorite

      Psalms he tells me.

I hope you like Hildegaard.

She is very nice and can she swim!

Sometimes she makes fun of my

      dogpaddle swimming - -

    but most of the time she

      is nice to me.

Of course I remind her that she

        can't

        BARK!

Grandpa Harvey says that GOD
has given each of HIS
creatures special gifts.

Do you know what special gifts GOD
has given to you?

Now I want you to meet "Birdie"
Her full name is
        Mrs. Birdie Blue.

You can tell by her picture
that Birdie has a big job!

In fact she is a bit tired.
If you look very carefully
at her eye you can see that
she is just a bit sad.

Sometimes we all can get
a bit sad as we get tired.

Birdie has just had a little
     talk with her family:

"You know,
     I work, work, work.
  I am up as soon as it
     is light

and I don't stop
    until it is nearly
    dark — in order to
    bring you food —
and all I hear is more!
            more! more!
It is enough to make
anyone discouraged!"

But.. Birdie loves her little
chicks ... and will continue to
work as hard as she can to care
for them.

Do you have someone that
loves you like that?

Someone who loves you
so very, very much, and
works very hard to help
you?

Please remember
to say "Thank you".

AND

"I LOVE YOU."

Now I want you to meet JIM

Jim is Grandpa Harvey's

youngest son.

Grandpa Harvey is a physician.

Jim was about 5 years old

when this picture was taken.

About one year later

Jim was very ill.

With the help of many prayers,

excellent nurses, physicians and

others of GOD'S helpers, and

penicillin, Jim made an

excellent recovery.

Jim's "TOOTH STORY" follows:

a note from

Jim's Dad —

This true story was written
in the summer following Jim's
illness. It is written in the
form of a devotional with
Scripture selections for each
of eight days. As you read
THE Scripture - day by day -
ask THE LORD to guide you
to HIS TRUTH.

Wm H Frazier, MD.

THE LORD AS HEALER

Let Us Pray: Our Father and Our God may you be glorified, as for the next eight days we prayerfully study a few of the many ways in which you "heal" our afflictions and distress, dispell darkness and gloom and break asunder. May we indeed give thanks to you, Lord, for your steadfast love, for your wonderful works to the sons of "men".   Amen.

A few months ago our son Jim, age six, suddenly was very ill. An infection began deep in the tissues of his face, distorting his nose and upper lip with swelling.

At medical school I had seen children with these fulminating infections in the "danger area" of the face. Penicillin was then unknown. The surgeons would try to find and drain any localized abscess so that the infection would not extend back along the blood vessels and into the brain. In spite of their efforts, most of these little patients died.

A dental x-ray showed that Jim had a large abscess at the roots of two teeth. The oral surgeon outlined the treatment: massive doses of penicillin, constant hot packs, force fluids--this to be done at home. Then "bring Jim to the hospital in the morning and after the penicillin and hot packs have done their work, we will operate, pull the two teeth and drain the abscess."

Jim was in pain and even with medication he could not sleep. To comfort him I told him stories. After arriving at the hospital and getting settled in bed, Jim became very quiet.

"Am I going to get well?" he suddenly asked.

"Yes, but I think it is going to take a little time." I replied.

"Daddy, my teeth hurt. Will you operate and make me well?"

On my finger I had a small cut. "Jim, will you heal this cut?"

"I can't."

"Nor can I heal you, Jim. It is the Lord who heals us."

"I see," said Jim, and gave me a little crooked smile.

GOD HEALS THROUGH HIS HELPERS

Let Us Pray: Our Father and Our God, we thank you for the gifts you give to man through your Holy Spirit. We thank you for giving knowledge and wisdom to your helpers--nurses, doctors, and scientists--that they may be partners with you in healing.   *Amen.*

"Is it really God who heals us?"

"Yes, Jim, it is. He works through His helpers--nurses, doctors, scientists." I was thinking of a plaque in the hospital library. "From God the doctor has his wisdom, thus God's creative work continues without cease."

All doctors are used by the Lord in healing. The Christian doctor knowingly participates in this ministry by praying for guidance and having prayer with his patients. This is no substitute for hard work or keeping up medically. It is a recognition that we physicians, of ourselves, heal no one.

"Will you operate on me, Dad?"

"No, the oral surgeon you met last night will be the Lord's helper. He will operate after the antibiotic does its work." I was thinking of a medical paper which began "From the earth shall come powerful medicines." A scientist was telling of his discovery of streptomycin, a powerful antibiotic. Strepto-- from the word streptococcus, a bacteria, and mycin--from the Greek word, "Mykes,' a fungus found in the soil of the earth. I was thinking of how penicillin was known for years as a mold which contaminated agar plates in a bacteriology laboratory. Then a scientist wondered if this mold which killed bacteria on agar plates could kill bacteria infecting a man's body.

A flash of genius---a research breakthrough---a creative thought---this I believe is what is meant by "Thus God's creative work continues without cease." I believe that just as God is the source of all healing, God is also the source of all true creative thought--in medicine as well as in music, art, or literature.

JESUS THE SHEPHERD

Let Us Pray: Our Father and Our God, we thank you for the Love you have for each of your children and for the many ways in which you show this love. Thank you for "wonder drugs" like penicillin.   *Amen.*

In Jim's room at the hospital was a picture showing Jesus holding a lamb in His arms. Jim loved the picture at first sight and asked me to tell him about it. Many times before had I seen this picture as I made hospital rounds attending patients, but not until noon of this same day did I begin to understand the picture.

At home before lunch, my wife read the eleventh verse of Isaiah 40--"He will gather the lambs in His arms---."

Returning to the hospital, I read this verse to Jim.

"Does Jesus love lambs too?" he asked.

"Yes, I'm sure He does. Do you know that He gave penicillin to lambs before He gave it to babies?"

"In the spring of the year shepherds take their sheep up into the high meadows in the Alps, the new grass is tender and abundant. There the baby lambs are born. Where ever there is 'good' in this world there is 'bad' also. Here the 'bad' is in the form of wolves. The shepherds and their dogs try hard to protect them, but every now and then a lamb will get hurt. The shepherds have known for centuries that if bread is kept in a warm, damp, dark place it will become covered with a mold. When this mold covered bread is placed next to a wound and then the wound wrapped, infection does not occur. Do you know the name of this mold? Penicillin!"

IN HIS IMAGE

Let Us Pray: Our Father and Our God, thank you for creating us in your own image. Thank you for sending Your son, Jesus, to tell us that by accepting Him as our Saviour we too, each of us, may become Your "adopted sons" with the spirit of Your Son in our hearts. (Galatians 4:4-6)   *Amen.*

"I don't get it" said Jim. "Does God love lambs more than He loves us?"

"No!"

"Then why did He give penicillin to lambs before He gave it to babies?"

"God gives many wonderful gifts to animals. He gave a sonar system to porpoises which our Navy would love to have. He gave a radar system to bats which our Air Force has not yet duplicated. But when God created man, he gave him the greatest gift of all. He created man in 'His Own Image'! Just think what it means, Jim, to be created in God's Image. It means that our mind is created in His Image too. I believe, Jim, that a man's limited (finite) mind may be tuned to God's limitless (infinite) mind and be in harmony. When that happens, that man becomes an open channel through whom God works---one of God's adopted sons with the Spirit of God's son, Jesus, within him."

Jim accepted the fact that he was made in God's Image with the trust of a little child.

How I wish more of my teenage patients could accept this fact. What purpose, dignity and healing is given to each human life when that person knows the answer to the question, "Who am I?" is---

A child of God created in His Own Image
--with a body to be kept as a sacred trust

--with a mind to be used as an open channel to express God's love
    and to continue His Creativity

--with a <u>soul</u> to glorify God and to enjoy Him forever.

PRAISE THE LORD

Let Us Pray: Our Father and Our God, we thank you that we may reach you by prayer and that through the power of the Holy Spirit you answer prayer.
Amen.

The nurse gave Jim his pre-operative medication and soon he became sleepy. He looked much better. The penicillin was doing a good job. Our minister told us that many people were remembering Jim and his surgeon in prayer, and then led Jim, his mother, and me in prayer just before Jim was taken to surgery.

The loving kindnesses of the Sisters and nurses to my wife and me during this time of waiting were most gratefully received. Our family called and expressed their love.

I had carefully anticipated the time Jim would be in surgery. Twice this amount of time passed before I became concerned. Then each additional five minutes seemed interminable. Sitting still was an impossibility. Finally the nursing supervisor came to relay the surgeons report--all had gone well; he would be down soon.

A nursing Sister said "Come doctor, get into this surgical gown and we will go up to the recovery room to see your son."

Just as we arrived, Jim raised his head and said, "It doesn't hurt now, Dad," his big smile showing two absent front teeth.

Simultaneously, the Sister and I quoted the 150 Psalm, "Praise the Lord---Praise Him for his mighty deeds---Praise Him with loud clashing cymbals." If I had had cymbals, believe me there would have been a loud clash!

In healing Jim, the Lord used His gift of tissue healing, His gift of penicillin, and His dedicated helper, the oral surgeon whose knowledge and training reflect an inheritance of medical knowledge given through brilliant minds of many nations over hundreds of years.

GOD IS LOVE

Let Us Pray: We thank you Our Father and Our God for the constant outpouring of Your Love upon each of your children. We thank you for the many ways in which you give us healing in body, mind and spirit.   *Amen.*

Psalm 107--"they cried to the Lord in their trouble--and He sent forth His Word and healed them." Luke 7:21--"In that hour Jesus--cured their diseases."

Today God continues to answer our prayers for healing. Sometimes He heals directly and instantaneously. Sometimes He heals in the ways presented in the true story we have been sharing.

God Heals--through His Love reflected in human love. Jim responded to the love of his family and his recovery was hastened.

God Heals--through the gift of "tissue healing" given when He created us. The cells of our tissues immediately start the "physiology of repair" whenever they are damaged.

God Heals--through His helpers in the profession of medicine and the healing arts. Their legacy of medical principals and surgical techniques came from the source of all creative thought--God Himself.

God Heals--and gives peace to our restless, seeking minds by giving answer to the questions "Who am I?" "What am I here for?" and "Who cares about me?"

God Heals--and gives serenity to our spirit by majestic natural beauty in mountains, lakes, trees and flowers.

God Heals--and gives refreshment to our spirit through inspired music, art, and literature. Hearing the "Sound of Music" will do for us today what hearing David's playing and singing did for King Saul. Through the artist, God can speak to the spirit of man.

These are a few of the ways in which God Heals us. All are evidence of His constant and everlasting Love. Every human being must have love to be normal and healthy. GOD IS LOVE.

JESUS CONQUERED DEATH

Let Us Pray: Our Father and Our God we thank you that you love us so much that through your Son- Jesus, you conquered death. Amen.

Man fears death. He is not born with this fear, but acquires it at an early age from his elders. He fears death for himself, and he fears death for his loved ones. UNTIL MAN IS UNSHACKLED FROM THAT FEAR OF DEATH HE CANNOT REALLY BE FREE TO LIVE. Man knows that Ec. 8:8 is true---"there is no man that hath power over the spirit to retain the spirit. Neither has he power in the day of death---".

When we pray to God for healing, does He always hear our prayer? Does He always say yes? In my practice as a physician I have learned to have prayer with a patient before surgery, asking His presence and healing touch upon the patient and guidance for me. Are all these patients healed? Is a patient found to have advanced cancer always healed? Does a baby born very prematurely always live? You know the answer.

Then God must sometimes not hear our prayers for healing or must sometimes say "no". I do not think so. I believe that GOD ALWAYS HEARS OUR PRAYERS FOR HEALING AND THAT HE ALWAYS SAYS YES. However, the time schedule and the locale for His healing is not specified. "The chief aim and purpose of man is to glorify God and to enjoy HIM forever." This statement from the Westminister Catechism has no geographic limitation either.

I believe that some patients have illnesses so severe and complicated that with our present limited medical and surgical knowledge they can be completely healed in body, mind and spirit only in Heaven. It is our limitation, not God's. For the Christian death is only a "calling to higher service"--a promotion. The day of death is "graduation day".

Limitations of space prevent me from sharing with you case histories of Christian patients who, though knowing they were dying, spent their last days on earth ministering to their families, friends and attendents. They were filled with a blessed peace and with a radiance which was very apparent to all who saw them. In the case of Marjorie, one doctor said, "What goes on here? I go in to attend the patient and it is I who comes away refreshed and inspired." Thus the dying of a Christian can be a Christian witness.

Love of God and surrender to Him does cast out fear. Death, the last enemy has been conquered--by Jesus Christ, and through Him by each of us who believes in Jesus.

Postscript — January 1995

Nearly 30 years have gone by since Jim's "Tooth story" was written in the summer of 1965. It appeared as a Devotional in August 1966 in "Today", a "Bimonthly Magazine for Home Worship" published by The Presbyterian Church. Jim is now an M.D. He is married and he and his wife Pattie, have three children Kelly, Andrew, and Stuart. The youngest, Stuart, is just two years old.

In 1988 three of my patients on the same day told me that they had been asking THE LORD in Prayer to guide me to write about how I began to have Prayer with my patients and to give them Scripture verses written on

my Prescription Pad.
   I took their request
seriously and asked THE LORD
to guide me in each detail.
When another of my patients
read Jim's Tooth Story in the
book "Scripture Prescriptions",
she asked me to put it in
a book for children.
   The Title " THE GIFT "
refers of course to the
Amazing gift that THE

CREATOR OF THE UNIVERSE
Would "so love the world
that HE gave HIS only SON,
that whoever believes in HIM
should not perish, but have
eternal life."
   What I wrote for
Day 7 of the devotional
      "GOD IS LOVE"
and What I wrote for Day 8
      "Jesus conquered Death"
I believe just as fervently

today as I did 30 years ago.
even more so !

Now that I am 78 years
of age and retired from
active practice I have more
time to seek THE LORD
early each day in Prayer,
and speak to others of HIS
children about HIS CONSTAN
LOVE !

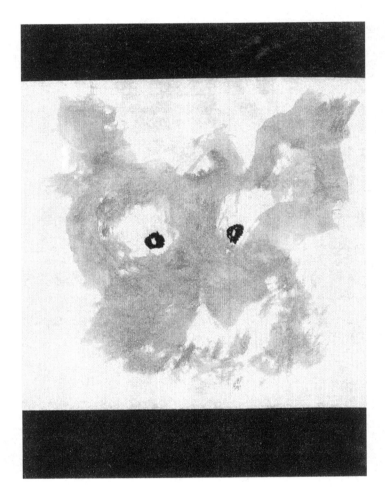

Hello Again—

I'm Back!

I told you at the beginning of this little book that I would like to hear from you---

You can write to me at the address on the next page — Zeke

Zeke
℅ Grandpa Harvey

PO Box 20026

Spokane, Wash.

99204—0026

The Science and Art of Medicine in Some of Their Aspects

by

Frederick Cheever Shattuck, MD

1847–1929

Jackson Professor Clinical Medicine, Harvard Medical School

1888–1912

Visiting Physician to the Massachusetts General Hospital

1886–19 12

SPAULDING-MOSS INC.

02090

Introductory Note

The timeless wisdom contained in this address by Dr. Shattuck so impressed the late Dr. Lewis Webb Hill and Dr. Grant V. Rodkey, staff visiting surgeon to the Massachusetts General Hospital and instructor in surgery, Harvard Medical School, that they felt it should be put in the hands of every medical student.

Accordingly, the address has been reproduced in pamphlet form at so low a cost as to permit of wide distribution by medical schools and other institutions.

The text is as it appeared in *A Memoir of Frederick Cheever Shattuck, MD* by George Cheever Shattuck, MD, which was circulated in 1968. During his lifetime, the impact of Dr. Shattuck's teaching and of his personality were felt and acknowledged from coast to coast in the US GCS.

The Science and Art of Medicine in Some of Their Aspects
by Frederick Cheever Shattuck, MD, 1969
*Address to the graduating class in medicine of
the medical department of Yale University, June 24,
1907. *The Boston Medical and Surgical Journal.*_Vol.
157 (July 11, 1907). pp. 63–67.

Medicine is truly a science in that it rests upon the foundation of facts ascertained or ascertainable. It is true that the fact of today may not be a fact of tomorrow. Were this, however, sufficient warrant for denying the term science to any branch of knowledge, whose withers would be unwrung? Whatever the discovery of radium may have brought, it has not brought peace of mind to the physicists and chemists. Medical science has grown marvelously of late in spite of complexity of the subjects involved and of the inherent difficulties which they offer to direct investigation. The more we know, the more we find we don't know. Science thus changes, and every great advance in knowledge is liable to require new hypotheses and theories which in turn prepare the way for the next advance. The medical knowledge of the public or laity is always that of the profession x years ago. We ourselves are responsible for the belief that pneumonia and pleurisy are due to exposure to cold and that uric acid is the cause of dire evils, including rheumatism.

The practice of medicine is the application of known facts and principles to disease. In preventive, the higher branch of medicine, this application is close. We acquire the power effectually to prevent malaria and yellow fever only when we learn how they are carried, though this knowledge has in no way helped us in the treatment of the individual cases. The same is true of typhoid fever. In the care and cure of diphtheria, on the other hand, science has enormously strengthened our hands, and we have confident reason to hope for similar aid in other infections. The surgeon's knife is in reality a confession of failure in so far as it is used for the relief of pathological surgery. Accident surgery can be abolished only by abolishing the activity, the life of man; but there seems no reason to despair of finding out how to prevent cancers and other new growths.

Those of us who are engaged in what is ordinarily termed the practice of medicine, that is, in dealing mainly with individual cases of disease, find that there is a gap between science and the concrete case. This gap is of varying width, here and there may be easily jumped, but must usually be bridged by the art of medicine. The practitioner of medicine soon learns that disease is one thing, the person diseased another. He finds, indeed, that in very many persons demanding his services, there is no disease in the sense which this word has come to convey, no definite and appreciable change in structure. Disease, discomfort there is due to impaired or disordered function, behind which may be a more or less obviously faulty mode of life, a false attitude toward life, or a mental or moral maladjustment to surroundings. The practitioner learns that human nature is essentially the same everywhere and in all times that it does not change as does the science with which he strives to cure or alleviate the ills of men. Osler tells us that in the last analysis, man has but two paramount desires—to get and to beget. The desire to get, however, is of wide extension, and as civilization advances, it is becoming ever wider, ranging from necessary food and protection against weather to automobiles and flying machines. But while it is true that human nature, in its main features, is unchanging, it is also clearly true and vital for the practitioner to realize that no two individuals are precisely alike in bodily and mental makeup and in environment. Each person is a kaleidoscopic figure, never exactly reproduced. Bertillon, whose name is perhaps most widely known through the application of his system of measurements to the identification of criminals, has shown that there is no duplication of size and proportion combined of that simple part of man's structure, his bony framework. Take a sufficient number of accurate measurements and you can pick your man from a multitude. Each person again has his own combination of lines on the palmar surface of his fingertips and hands. If there is such infinite variation in the simple, is it not reasonable to suppose it to exist in the complex, in internal organs and their balance, above all in the ganglionic nervous system which should ensure the harmonious working together of the various organs and other systems? As regards to the brain, indeed, there can be no question, so open to observation are its workings and so great is its capacity for education.

It is thus clear that our work is not mathematical, nor is there any danger of its becoming a penny-in-the-slot business, certainly within a future which can be of other than speculative interest to any of us here present. Science has its pure devotees—biologists, physiologists, and pathologists. The art of medicine also has its pure devotees, the quacks. I assume that very few of you here presently aim at the former, none at the latter. And yet there are quacks within and on the outskirts of the profession, the worst ones, for they have known light and have deliberately chosen darkness. I have known highly trained men who have gone as far professionally wrong as a man can go, bartering their birthright for a mess of pottage.

Most of you, doubtless, look forward to being practitioners. For four years now, you have devoted yourselves mainly to the science of medicine, to its fundamental facts and principles, and to the methods which we employ in trying to solve the riddles of disease. Much that you have learned, you will have to unlearn. How brief is the life of a textbook on the practice of medicine! Many books which well served their day and were constantly in the hands of the progressive practitioner now gathering dust on the shelves of libraries. Here and there one, Watson's *Practice*,[1] for instance, has an occasional reader attracted by its style. We think we are too busy to read an old book, but when we do so, we find that the masters knew their art, though the young student of today knows more science. He who reads, marks, learns, and inwardly digests the chapter on dyspepsia in Austin Flint's *Practice of Medicine,* [2] to make a random illustration, will be more successful in his treatment than the man who relies alone on the latest monographs on diseases of the stomach and guides his therapy overmuch by the stomach tube and chemistry.

You will now begin to apply your science, that is, combine science and art with direct responsibility—that great teacher—all the time keeping abreast with the advance in science, entering the post graduate school which claims you as long as life lasts.

The art of medicine is not easy to teach directly. Something of it you learn more or less unconsciously in watching your clinical teachers and their handling of cases—something negative perhaps as well as positive, something to avoid as well as to imitate. If difficult to teach, there is much to learn, and it may be that thirty-odd years of practice and nearly thirty years trying to teach medical students give some warrant for offering you a few hints of possible use.

Success in the practice of medicine lies in a happy marriage between science and art. It is as sure as success in any other calling, though there are no great money prizes. He who gauges his success in life by the amount of money he makes had better resign medicine to those who seek happiness and a reasonable support in the exercise of their faculties in a calling which is ever growing, is of prime service to man, and gives a chance for unlimited self-development. It means hard work, but you remember that Sir Andrew Clark[3] defined work as the life of life.

The elements of success seem to me somewhat as follows:

1. Of course, you must know your business and possess a reasonable knowledge of your profession as a science. This, with due diligence in its exercise, the law demands. The defendant in a malpractice suit is not expected to know all there is to be known on the subject. Indeed, that wise physician, Dr. James Jackson,[4] said sixty years ago that there was at that time more known about medicine than the mind of any one man could grasp. What would he say today? The adjective reasonable is elastic and must be well-stretched from the legal limit to satisfy the demands of conscience and to attain real success.

2. Thoroughness. Form the habit of it from the start and deal with each case which comes to you as if it were the only case in the world. Never forget that mistakes come far more from not looking than from not knowing and from taking things for granted. Collect all your facts. For failure to do this, there is no excuse. All can do it.

The evidence which leads to diagnosis and treatment is of two classes: the objective, which you gather for yourself through observation and physical examination and the subjective, which you elicit by questioning the patient.

215

Of the two, the former is by far the more valuable and can never be dispensed with. But we need the latter and must then often call in the aid of time before we can reach a definite conclusion. Sometimes patients get well without our really knowing what was the matter with them. Sometimes the autopsy fails to clear up the mystery. I wonder whether we are as keen observers as were our forbears who had no instruments of precision to rely on, who knew nothing of germs and their ways.

There are cases in which time is saved by making the physical examination before taking the history, though I am inclined to think this order is safer for the experienced than for the younger practitioner. If there be very definite organic disease, it does not, as a rule, escape searching exploration. And yet we elders have all seen cases enough which we would pass as good risks for life insurance if we relied on the objective evidence alone. Perhaps angina pectoris affords as good an illustration as any. The heart, arteries, and kidneys may seem perfectly sound while the age of the patient and his symptoms reveal coronary change which may indeed soon prove fatal in spite of the best care, which under mismanagement is likely to prove very serious but which so often is amenable to wise counsel persistently and faithfully followed out. The sequence and development of the symptoms, their mode of onset, may be essential to the determination of their exact pathological nature, the sudden or slow advent of paralysis, for instance. The seat of a lesion may be revealed by physical examination alone while the nature of a lesion can only be determined by the history. Even when physical examination fully lies bare the disease, it may tell us little as to the subject of the disease, and our advice must be given to the individual, not to the average man.

In history taking, there is a chance for the exercise of much art. The way in which a person tells his story in itself teaches you much about his case. The talk of some people, not always females, like the brook runs on forever, though unlike it, gets nowhere. Others seem unable to give a direct answer to a direct question. Important facts may be forgotten and may be suppressed. Some are so foolish as to deliberately deceive or attempt to deceive their doctor. There are cases in which it is best to let the patient tell his story in his own way. Others in which question and answer should be resorted to, either from the start or later. Just remember that people generally care little how you collect your facts. They want to help you to help them and are ready to accept your methods, especially if tactfully applied.

It is interesting to recall that the prototype of Sherlock Holmes was Mr. Bell of Edinburgh, one of Conan Doyle's surgical teachers. The power of observation is capable of great training. Don't think thoroughness a waste of time. You will find that, gradually, you work quicker, though no less surely, lower centers learning how to do that which at first only the higher could perform.

3. Common sense may for our purpose be defined as the power to arrange all the facts with an eye to perspective, with a sense of "values" which puts into the foreground those facts which are essential, properly groups those which are secondary and rejects those which are irrelevant to the case in hand. It includes the power to see what is possible and practicable in any given case. The ideal is rarely attainable, and we too often have to be content with as close an approximation as may be possible. Perhaps common sense may be made to include the power of logical reasoning. Here there may be excuse for failure such as there is not for lack of thoroughness. Clear vision and an orderly mind are not given to all in the same degree, but I am sure that they are cultivable, and the problem for each of us is to make the most of himself.

4. Character, though hard to define, is very real and a very important factor in successful practice. Character is elevated by the formation of high ideals and strengthened by the resolute pursuit of them, the surmounting of obstacles internal and external. Character seems to be sometimes latent and comes out only under a squeezing process which may prove to be of incalculable benefit although painful at the time. Character will invariably reveal itself. Do not be studied or self-conscious. Do not ask yourself, "What effect am I producing?" A man should be himself, eradicating and restraining his bad, cultivating his good qualities. Do your work the very best way you know how, and the result is sure to follow. The character of the physician has much to do with the obedience of the patient and with putting him into the frame of mind which leads to cooperation.

5. Enthusiasm and a genuine love for his profession, apart from its emoluments, is a quality of great value to the physician. In no other calling is there greater justification for enthusiasm. It is our privilege to try to promote the well-being of man, and we cannot do our best and highest work unless we are ever mindful of the relation of mental and moral states to ills either arising in the body or outwardly manifested by it. A love for your work will support you in discouragements of all kinds which in greater or less measure must come to all of us in life. Drudgery cannot be escaped in any calling. It is part of life, but there is no drudgery in any way comparable to that endured by the man whose sole aim is to amuse himself.

    I remember hearing one of my teachers in college days, who might then with profit to all concerned have availed himself of the Carnegie Pension Fund had it existed at the time, remarking in the course of a lecture, "Gentlemen, you may think what you will, but you will find that whatever you do in life it will come down to turning the crank." God help the man of medicine, and still more his patient, who believes or gets to believing this to be true.

6. Next may come sympathy. The power of putting yourself in another's place and of realizing how he feels under given circumstances. Try to catch his point of view. If you cannot do this, act on what you think would be your own in the same situation, remembering all the time that his and yours are not necessarily the same. One personal lesson on this point is so deeply stamped on my memory that I trust you will pardon its mention. I had been seeing daily in consultation a gentleman who had special reasons for wanting to get well. He seemed on the road to recovery, and while returning to the station with his doctor, I suggested that it might be well to skip a day, that the moral effect of so doing might be good, the fact of omitting a visit strengthening our assertion that we thought him doing well. My colleague assented. About bedtime that night, I got a telegram. "Mr. A wishes you to come down as usual tomorrow." When the doctor met me the next day, he reminded me of our conversation of the previous day and of my using the words "moral effect." He then went on to state that when he told Mr. A that no consultation would be held the next day, the patient remarked, "I think he had better come—for the moral effect!"

A uniformly kind manner, especially if it be based on kindness of heart, counts for much. Abernethy's[5] success did not depend on his brutality. I like to recall the saying of an old French physician with whom I was visiting at the Hotel Dieu at Lyons, *"Le Mdecin gurit rarement, ii amliore souvent, ii console toujours."*

Cheerfulness in the physician is a good tonic to the patient. Our fears are for ourselves and for a judicious friend or member of the family, if haply such there be. Unreasonableness on the part of a sick person is not to be regarded as it might be where he well. Illness throws some of us off our balance; but the lessons in fine courage, constancy, unselfishness, which we may all learn from our patients, are far from infrequent. The juice of character flows under the press of adversity.

Remember that the doctor's visit should be the main incident of the day to the sick man and try to make it as near the same or expected hour as you can consistently with other duties.

Once again, do not confound the sick man and his disease. Laborious days and broken nights are compensated in part by the fact that no one gets so close to the real man as the doctor. The lawyer gets closer to his money, which some may value more than themselves. The clergyman may be liable to get closer to the would-be than the real self, to the John's John rather than to the real John, as the wholly delightful *Autocrat of the Breakfast Table*[6] puts it. The doctor sees plenty of evil, more weakness, but still more of good in human nature. While the head must rule the heart, the latter cannot go untouched.

7. Honesty of mind is not, it seems to me, quite the same thing as common sense. Its essential feature is the power of seeing things just as they are without being blinded by either prejudice or personal interest. In the higher degrees, it is given to few. It is to be sharply differentiated from honesty of act, though when the two honesties are combined, they are precious alike to the owner and his circle, be it large or small. The man with an honest mind may act dishonestly, with wide-open eyes doing that which he knows to be wrong. His vision is clear, but his character or standard is low. On the other hand, the man of dishonest mind may act honestly and be governed by only the highest and most unselfish motives. The muddle-headed good man can do great harm and may even bring goodness into disrepute.

We can perhaps as well here as anywhere touch on the devotee of pure art in medicine—the quack who certainly scores his successes and thus becomes worthy of analysis.

Is not the essence of quackery the claiming to possess power or knowledge which the claimant knows he does not possess? Thus, dishonesty is inherent and implied. The visionary, the fanatic, or the ignorant man has deceived himself before he deceives others. He is a crank rather than a quack. But he may be both, in that he may gradually have come from clear to troubled vision, to self-deceit, a thing against which we must everyone of us be constantly on our guard, so largely does the ego enter into our composition. The quack generally knows his human nature and the salient fact in medicine that the tendency of most pathological processes is toward recovery. He is a past master in the use of suggestion, through which the mind influences the body, the vital element in much drug action and in all forms of irregular healing. He does not hesitate to promise results which we, handicapped by knowledge of our ignorance and by conscience, often cannot do. We are expected to cure to a degree that he is not. It therefore often happens that it is our failures, his successes, which are spread abroad. Those who have resorted to him and have failed to get the promised relief are generally ashamed of their foolishness and keep quiet. Our failures, mistakes, or supposed mistakes, on the other hand, become the common property of the sewing circle and may travel far by the most up-to-date means of communication. The way of the fashionable practitioner, whether he transgress or not, is not always easy.

The doctor is judged by a more or less incompetent tribunal, the proportion of people who know what constitutes evidence in a question of physical science being small. There are not as many "born doctors" as there used to be, and the seventh son of the seventh son makes less noise in the land, though his trumpet is still audible. Let us not be distressed by seeing the quack flourish near us, even if it be as the green bay tree. He, like the dog, has his day, usually a relatively short one. Neither should we let our minds be troubled by criticism, however undeserved, provided that we have the approval of our own consciences. Let us never forget that we receive far more credit which we do not deserve than we do discredit. Let us take the lean with the fat without grimace. We can recall the saying attributed to Mr. Lincoln about fooling all the people all the time. I say attributed because an earnest searcher told me the other day that he had been unable to find the saying anywhere in Mr. Lincoln's writings or to find out when and where it was said.

With the above qualities kept constantly under cultivation, good health and good habits, you are bound to succeed sooner or later, more or less rapidly. Your lives of useful service will be happy and the ranks of those who beg their bread have very few recruits from the children of the righteous professional man.

The essential aims of undergraduate medical teaching seem to me only three: first to teach the fundamental facts, the doing which must open before you more or less clearly the lines of future progress; second, the methods to be employed in the study and recognition of disease; third, thoroughness in work. Between the latter and the Yale spirit there is, I have reason to believe, a peculiarly close sympathy. Thoroughness is a close link between the science and the art of medicine, which latter may be summed up in the observance of the Golden Rule. Thoroughness too is a prophylactic against routine, the favorite child of sloth. Rules are made to break. They are our servants and should never be allowed to become our masters. We make rules to save thought, the hardest work there is. We must think whether and how far the rule applies to the given case, and let us not hesitate to break it for cause.

There are other points I would fain touch on, but I will be merciful. A sermon, from which these remarks are not wholly devoid of resemblance, never used to be complete without a lastly and a finally. Under, lastly, a few words on the point of ethics as applied to the practice of medicine: Are we ever justified in deceiving or lying to our patient? My answer to that question is an unhesitating "yes," but—that little though significant word forces itself forward—only under conditions. Before stating my view of them, let me ask anew the question of Pilate, "What is truth?"

In matters medical, we are dealing with x, i, and z, and we must never forget that our opinion does not necessarily represent the truth. The truth, as we see it, may not be the truth itself. In telling the truth as the truth, we should be very sure that it is the truth, especially if in so doing, we condemn the patient to an incurable and painful malady. There is generally a near friend or relative to whom we can speak frankly if we like, expressing our fears from which we should not be too ready to take counsel. But apart from our liability to error, it seems to me that every practitioner now and then meets a case in which neither silence nor skillful evasion avail, and in which telling the truth is sheer cruelty. The cases are not common in which a categorical lie is justified or necessary, but there are cases in which the question "Have I a cancer?" should, in my opinion, be answered by a firm "No." There are, again, cases where the knowledge of alues would break up a family and wreck the happiness of the innocent without doing them any good. There would be no difficulty in multiplying illustrations. Doubtless, ordinarily, the truth, which may be stated with brutal frankness or delicate consideration, is best for all concerned, especially if it be served with the saving grace of "as I see it," or its equivalent.

Now for the condition or touchstone, before swerving from or denying the truth, we should ask ourselves the searching question, "For whose advantage is this denial?" If it is in any measure for our advantage or seeming advantage, let us shame the devil. Only when in our deliberate judgment the interest of the patient and his family, and that alone, is thereby furthered is the truth to be denied.

Finally, allow me, my younger brothers in the profession, to wish you Godspeed and the attainment of your heart's desire.

Notes to the Text

The Science and Art of Medicine in Some of Their Aspects

1. Thomas Watson (1792–1882), *Lectures on the Principles and Practice of Physic,* was published in 1843. It remained the chief English medical textbook for the next three decades.
2. Austin Flint (1812–86), MD, *A Treatise on the Principles and Practice of Medicine,* Harvard 1833, founded the Buffalo Medical College (1847) and helped found Bellevue Hospital Medical College, New York (1861). His classic work was originally published in 1866, was many times reissued and revised.
3. Andrew Clark (1826-93), born in Aberdeen, Scotland, was physician to the London Hospital from 18661886.
4. James Jackson (1777–1867), MB, Harvard 1802, and MD, Harvard 1809, was Hersey professor of the Theory and Practice of Physic at the Harvard Medical School (1812–36). He was also a founder of the Massachusetts General Hospital (1821) and a Harvard overseer. See page 278 of *A Memoir of Frederick Cheever Shattuck, MD* for FCS's account of him.
5. John Abernethy (1764–1831) was noted London surgeon and lecturer on anatomy and physiology, FRS, 1796. Several editions of his *Memoirs* edited by George Macilwain, MD have been published.
6. 6. Oliver Wendell Holmes, "The Autocrat of the Breakfast Table" was a famous series of essays contributed to the first twelve issues of the *Atlantic Monthly* founded in 1857 and later published in book form (1858).

Offs.tPmnedby

SPAULDING-MOSS INC

W.shd. Mo,sch!.ts 01090

Index